D0631854

Sleep Talkin' Man

Sleep Talkin' Man

Karen Slavick-Lennard

Illustrated by Tom Daly

BLACK CAT

a paperback original imprint of Grove/Atlantic, Inc.

New York

Printed in the United States of America

ISBN: 978-0-8021-7093-4

BLACK CAT
a paperback original imprint of Grove/Atlantic, Inc.
841 Broadway
New York, NY 10003

Distributed by Publishers Group West
www.groveatlantic.com

13 14 15   10 9 8 7 6 5 4 3 2 1

"This is **my** story. It starts with **me**. And it ends with **me**. And everything in the middle is **about me**. Greatest fucking story ever written."

MY HUSBAND IS A LOVELY PERSON.

I feel compelled to repeat these words every time I talk to someone about, or meet someone who has heard about, Adam's sleep talking. You see, about three-and-a-half years ago, I suddenly discovered that the sweet, mild-mannered Englishman with whom I share my life has a very different side to him that is only revealed when he goes to sleep. By day, Adam is delightful: kind, generous, clever, romantic, insightful, humble, respectful, infinitely likeable. You'd love him. But at night, the man who chatters at my side is ruthlessly cutting, profoundly egomaniacal, bizarrely perverse. In short, the antithesis of his waking self.

So before you go poking around in this book, please take my word for it that Sleep Talkin' Man—the alter ego, the insult comic, the baby-juggling donkey jockey—is an aspect of my husband that he reserves entirely for his sleeping hours. Adam does not wish to exterminate all vegetarians, he does not spend his days lobbing the c-bomb at everyone in his sphere, and he does not have a badger trained to attack on command (although, that could be pretty handy sometimes).

So what is Sleep Talkin' Man? Where does he come from? Is he a monster comprised of the darkest elements of Adam's soul, kept repressed and hidden from the waking world? Is he the spectre of obnoxious, self-obsessed men everywhere crying out to be heard? Is he the cynical, embittered spirit of Charles Bukowski? We'll probably never know. Best to just enjoy him while he's around, and be thankful he's not running the nation.

As far as anyone knows, Adam had never really talked in his sleep before I came along. His parents had never heard him. His former wife of eight years never heard him. He does have an old girlfriend from university who came forward and reported that Adam used to randomly swear in his sleep now and then, but nothing more elaborate than that. So it would appear that Sleep Talkin' Man was born late one February night in 2009.

The end of that evening had gone the way of most of our evenings: Adam had given me a kiss goodnight, snuggled down into covers, and was asleep within ten seconds. I had gazed down at him in envy, sighed, and turned my attention to an online Killer Sudoku puzzle, hoping it wasn't going to be another one of my insomniacal nights.

About half an hour had passed like that. I was still doing my puzzle, waiting for sleepiness to steal over me. The flat was silent except for Adam's steady rhythmic breathing and the ticking of the kitchen clock. Suddenly, Adam reached up, patted my head and intoned, "Yeeeesssss. Sleeeeeeep." OK, perhaps this was a bit peculiar. But it was relevant, and good advice, so I didn't think much of it.

A few minutes later, he declared quite emphatically into the silence, "Enough with the cheese. Enough!" Well, sure, I do eat a lot of cheese. I'm not ashamed to admit it. But my cheese consumption hardly seemed worthy of such scorn. I started to giggle, and then to laugh out loud. Not surprisingly, my guffaws roused him from his mumbling slumber. When I relayed what had happened, I'm not sure that he actually believed me, but he giggled along anyway.

It was a few weeks before I heard from Adam's sleeping self again. "Little people are FUNNY!" he muttered in my ear during an afternoon nap. Yikes, I thought, I guess my husband is not quite as politically correct in his sleep as he is in his waking life.

It was in mid-July though, still very early days, when Sleep Talkin' Man fully revealed his true colors. It was a peaceful, silent night, when out of the darkness I heard, "Consider yourself fired . . . dickhead." Whoa. This was the first indication that I was spending my nights with a truly different animal than the man I had married. First of all, I'm pretty confident that, until that moment, Adam had never uttered the word "dickhead" in his life. Not because he's averse to swearing when the situation calls for it, but as an Englishman from an upper–middle class background, he'd be far more inclined to use the traditional "wanker" or "tosser" when out to demean. And on top of that, I couldn't imagine Adam could ever sack someone with such obnoxious yet effortless aplomb. That was when I started to think of Adam's sleep talking as the words of not the Adam I knew, but of someone else entirely. And he could be a real, well, dickhead.

From that point on, Adam talked in his sleep more and more frequently. It started out every couple of weeks. Then it was every few nights. Then nearly every night. And finally, loads every single night. Sleep Talkin' Man came to have such a strong, individual presence in our life, we began to think of him as an entirely separate entity, a sort of third spouse who, instead of helping around the house or earning money for the family, provides us with regular, if unusual, comic relief.

Sleep Talkin' Man's appearance in my life also

kicked off an eye-opening education in British slang. Never from Adam alone would I have been exposed to the infinite ways in which the English can employ our shared mother tongue to patronize and humiliate. The "C word" alone—used much more liberally in the United Kingdom, and generally as an insult to men rather than women—has been demonstrated by STM (as we call him for short) to have inexhaustible application.

(When putting this book together, we had to decide whether to leave the words exactly as STM spoke them, or make them more American-friendly. We settled for keeping the language intact, although we did convert to American spellings. So as you read, keep in mind, my fellow Americans, that "pissed" means drunk, "knickers" are panties, "cunt" is not as bad as you're used to, and if you come across a completely bewildering word or phrase, well, it's probably one of those silly British things. I've included a handy glossary of Britishisms in the back of the book, for those occasions when context just isn't enough.)

From the beginning, I was enjoying Adam's sleep talking far too much to keep it to myself. Skype sessions with my friends and family back in the United States always began with an update of the latest and greatest from STM. Meeting new people, I always found a way to work the subject in at parties, at the grocery store, to anyone who'd listen. I just couldn't get enough of it.

I loved Adam's sleep talking so much, I decided to share it with the world, via a blog. And now, it's come to you in this book. I hope you get even a fraction of the delight reading STM's words of wisdom that I got from hearing them uttered in the night. Or, if nothing else, I hope you come away with a truly shattering insult to use in just the right moment, awake or asleep.

"

I'm so happy I could shit a puppy.

●

So this is what it feels like to be a gummy
bear! I can't walk though, I have to rock.
I think I'll call myself BerNARD. Not BERnard.
BerNARD. And I'll be a golden gummy bear.

●

Even my reflection is jealous of me.

●

You know, it would be easier to kill the
vegetarians than invite them to dinner.
Bloody fucking lentils.

●

Congratulations. You may now wipe
your face on my butt cheeks.

"

"Well, fuck you! Fuck you, Amazon. You don't KNOW me. I kill puppies. Suggest something for me NOW then, you shit-spreading bastard—Ooh! Bathtub fairy lights! Mmm, pretty.

❧

You really are life's wet patch. An embarrassing little stain that no one wants to admit to . . . or sleep on.

❧

I'm sorry. I'm gonna have to bag up all my nasal hair. It'll take me hours, but, it will be worth it! Breathe in, breathe out. . . . I didn't fucking say breathe in again!

❧

I've got a great game. You get a car, and a house. A good life. Then shit happens, you lose everything. I'm gonna call it Grand Theft Divorce."

“

A tense moment at the farmyard roller disco competition . . . Everyone's waiting for chicken to get her skates on . . . Will she perform, or will she chicken out?

❦

I shit gold, piss silver, and puke bronze. I don't need a medal to tell me how fucking awesome I am. Got that, bitches?

❦

Go on, cry motherfucker. I'm gonna dip my dim sum in your tears.

❦

Watch out! I've got a hot-pant leg-warmer moment coming on.

”

"

If at first you don't find love . . .
Google harder!

*

You know what's missing from your pictures?
Artistic fucking talent. Now get a proper job,
you loser.

*

I'd rather peel off my skin and bathe my
weeping raw flesh in a bath of vinegar than
spend any time with you. But that's just
my opinion. Don't take it personally.

*

Handle with care: contains awesomeness.

*

You really are two I-don't-cares
short of a fuck-off.

"

"

I hope you guys don't mind,
but I brought my vegetarian substitute.
Would you mind cooking that roast
leg of lamb for me? Just don't let your
veggie burgers touch it. Thank you.

❦

Badger tickling: proceed with caution.

❦

There's a little bit of me that's just a
little bit excited. A little fluffy bit.
No, you CAN'T SEE IT!

❦

I am simply far too busy being passive
aggressive to give a shit about you.

❦

Yeah, you can find it on
www.uselessfuckingpieceofcockshitesuckingtitfuck.com.

"

"

You're gonna have to shave your pubes. It's like fighting an army of permed spider legs down there, and I'm gonna lose. I'm gonna lose.

♥

I'm in a totally zen-like state . . . Can't you tell how fucking zen I am?!

♥

You got an issue with my goat, you got an issue with me. Come on, goat. We're going somewhere where we're welcome. Baaaaah.

♥

Seriously, what am I going to do with a dead rat? Honestly. Your presents are getting worse and worse.

"

"

**Fuck it.
I'm gonna build a fetish costume
for my chicken.**

"

> "Shhhhhhhhh. I'm telling you: Your voice, my ears. A bad combination."

You may be wondering how I get any sleep with Adam chattering the night away. In truth, it's not as bad as it sounds: First of all, Adam does all of his talking for a night in one stint. Even if he comes out with, say, six quotes, it's all within a fifteen-minute period. So, it's not as if I'm being awakened every hour by his prattling on. And then there's my chronic insomnia. Often when STM makes his appearance, I have been lying there wide-eyed and alert for a while, desperate for any distraction. On such nights, Adam's sleep talking is a gift that breaks the monotony of those long sleepless hours (which may help to answer the question of how I can be so amused by STM, instead of wanting to, say, slap some duct tape over his mouth). Of course, there are occasions that his sleep talking does wake me from a sound sleep. But by now, I am as attuned to it as a mother to the late-night cries of a child; no matter how deeply I am under, when STM appears I immediately snap awake. There is no practical reason for this, mind you, as I have the recorder going all night. But I love hearing STM's pronouncements first hand, and besides, if I'm awake, I can hold the recorder right near his head for maximum audio quality. Conveniently, Adam's sleep talking sessions are always preceded by a couple of quiet little groans or whimpers, which immediately wake me up and give me the five seconds I need to achieve full alertness, grab the recorder, and assume the position. Amazingly, the loudest snoring from either Adam or Molly the little beagle doesn't disturb my slumber in the slightest, and yet the tiniest "hrmph" has me instantly primed and ready for action!

I know what you're thinking: How did a nice girl like me end up with a digital recorder permanently installed on the bedside table? Well, when Adam first started talking in his sleep, it was so novel and surprising for me that those early quotes were stamped indelibly on my memory. After those first few times, though, it was clear that I soon wouldn't be able to keep it all in my head anymore, so I started keeping a written log for myself. Over time, Adam became more prolific, offering a number of quotes in one night, and I could no longer trust my memory to make it until morning. So, the moment he would start jabbering, I'd immediately pop open the computer and start typing away furiously, trying to keep up with his quick-fire pace. This approach had a number of issues:

1. It was stressful. You try taking dictation at top speed in the dead of night!
2. As surreptitiously as I tried to type, Adam would start incorporating the sounds of the typing into his dreams (clip-clop of horses, applause, rain). This was amusing, but always ultimately resulted in him waking up.
3. I was reluctant to go to sleep AT ALL, in case I missed anything.

Once I took note of that last one, I knew something had to change. And that's how I ended up with my trusty little Olympus. Now I press the red button every night before we go to sleep, and snuggle down for as a good a night sleep as my chronic insomnia will allow. I still hear everything STM comes out with first hand, because Adam's sleep talking invariably wakes me, but now I can just lay back, stress-free, and giggle into my pillow.

For the geeks out there—or otherwise interested folks—here's how it works. First of all, the recorder is sound-activated. So I'm not faced with eight hours of oblivion to go through every morning. I end up with somewhere between thirty to sixty minutes. Of course, most of that is not sleep talking; it's Adam and me having a final chat, the heater kicking on, Molly snoring loudly, the subtle creak of the bed every time one of us turns over, and a host of other nighttime noises. I certainly don't have the time or patience to sit and listen through all that once a day. Instead, I LOOK through the recording. I open the mp3 file in music editing software, and look through the waves to find the sleep talking. After three years of doing this, I can identify with barely a glance the pattern of a throat clear, a snore, a sigh, a truck rumbling past. . . . So, amongst all of that, I find the actual sleep-talking quotes and create separate little audio clips of each. At the moment that I'm writing this, I've got 1,735 of them.

Having a recorder going from bedtime until morning every night has a further benefit, one that has nothing to do with Adam's sleep talking: it captures many lovely little moments in our relationship. I'm a hoarder of sentimental crap of all sorts, and my inability to delete audio files of our late-night and early-morning conversations is no different. In fact, I even have a recording of a conversation of us talking about me having recordings of our conversations!

KAREN: *You know what's nice? If you die, I'll have all these recordings of us talking.*

ADAM: What? No! "Do you know what would be nice? If you die..."

KAREN: *No, that's not the part that's nice—*

ADAM: Imagine if your tape recorder shut off right then.

KAREN: *Let me try it differently: If you die, you know what would be . . . nice? Even though that's horrible, that you're dead, what would be nice apart from that? That I'd have all these recordings of us talking.*

ADAM: Listen: "I just thought of something really wonderful, which is no matter what happens to you, I'll always have recordings of us talking."

KAREN: *Yeah, that's a better job.*

ADAM: "You know what would be nice? If you die . . ." You wake me up, to tell me it would be nice if I died.

KAREN: *But you know what will be really ironic, is listening to this conversation then.*

ADAM: You'll be like, "Yeah, this really IS nice!"

"

Good morning.
I just wanted to be the first person
to call you a twat. Enjoy your day.

Oooh. You know what would be
good? Ass bags. Big bags full of ass.

You know, you're not some precious
flower. And if you were I'd be a weed
and grow next to you and CHOKE YOU
TO FUCKING DEATH! . . . Love you.

Don't worry. I'll find it. That's what I do,
find things. I find you annoying. See?

"

"

Fuck! If I don't get to
the motherfucking flower show,
I'm gonna fucking kill someone!

❡

You're so vain, you probably think
even the mannequins are checking you out.
Idiot.

❡

Don't worry. If you kick one bunny, then all
the rest will scatter.

❡

Hurray for me! Yup, I said it. Now
the rest of you can join in.

"

"

Oh, there are some noises that really bug
the shit out of me. Your talking . . . and
your breathing.

❦

Why don't you call back later,
and we'll see whether we can get the world
to revolve around you.

❦

Of course the zombie loved me.
She gave me her heart. Mmmmm-hmmm.
And her hand in marriage.

❦

Great. So you've fallen in love again.
Go wash yourself off. Dirty stuff.

"

"

Skipping to work makes everything better.

♥

Yes, I can get away with wearing leather chaps. Just not on a windy day.

♥

It's time to don our cow masks and scare the salad out of her. I love Veggieween.

♥

Even your reflection thinks you're a pathetic piece of scum. Now leave me be. I've got some breathing to do.

♥

My pony's for sale. . . .
Yes, it fucking works!
I wanna get a stripey zebra instead.

"

"

I'm gonna puke in your pants.
I'm gonna be the knicker puker. Yeah.

♥

You know, with you you you, it's all me me
me. Well fuck fuck fuck fuck you you you.

♥

Can you hold my starfish? It doesn't like it
when I'm getting excited. Oh look,
it likes you! Its legs are all creepy creepy.

♥

I can't control the kittens.
Too many whiskers!
Too many whiskers!

"

“

Hands up who wants sweets!
Hands up! Awwwwww, you've got no hands.
Sucks to be you, sweets for me.

♥

Pygmy goat herding sucks. I've got this
fucking teacup pig for a ride, and they just
all laugh at me. Fucking mini-bearded
bitches.

♥

With just a little perspective, you will
agree that you're a festered limp fucking
dick. Shrivel up my friend, shrivel up.

♥

I'm gonna have a great day . . .
Don't you fuck it up.

”

"

I have succumbed to temptation! Mankini or body wax? Decisions, decisions . . .

♥

I've got something for you. It's a future. And you can have it when you leave.

♥

Oh, I hope you take this advice to heart: you look fat when you cry.

♥

It's taken me years to get things just fucking right, and one monkey comes along and ruins everything! Stupid monkey! Grrrr.

"

This little tampon went out, this little tampon stayed home. This little tampon had an applicator, this little tampon had none. This little tampon's covered in . . . poop. WRONG HOLE, PEOPLE! Wrong hole.

Yeah, keep laughing. Laugh your fucking face off, you giggling little freak.

> "It's definitely time to get up.
> Yes. My dog needs a new tutu."

Adam's bouts of sleep talking often culminate in a spontaneous awakening. Sometimes those awakenings are mild but odd, with a sudden intake of breath and an immediate declaration of some tidbit of information that, considering that he has just emerged from sleep, turns out to be surprisingly relevant to the day; more often, his awakenings are dramatic affairs: a exclamation yelled into the silence or a fist to the headboard. In either case, Adam is instantly alert.

Since he often wakes up to find me giggling next to him, Adam gets just as excited to hear what he has said in his sleep as I am to tell him. Our ensuing conversations—dubbed "reveals" by me—can get pretty wacky. Luckily for me, the recorder is generally still going, allowing me to chuckle over them later.

Sometimes Awake Adam carries on with STM's flight of fancy. For example:

> "Don't let the midget out of the wardrobe. No! He doesn't come out until Thursday . . . Not until Thursday."

ME: *You said, "Don't let the midget out of the wardrobe. He doesn't come out until Thursday."*

ADAM: He's got to polish all my shoes. Cleans my shoes, straightens my shirts, and guards against moths. Thursday is his day off.

ME: *Are moths afraid of midgets?*

ADAM: He eats the moths.

ME: *Oh. Does he get to eat anything else?*

ADAM: Moths and dust.

ME: *Dust is mostly human skin cells.*

ADAM: He eats dust.

ME: *So, you're saying he's a cannibal. Aren't you afraid of keeping a cannibal around the house?*

ADAM: No, I don't keep him around the house. I keep him in the wardrobe.

ME: *Where does he go on Thursday?*

ADAM: I let him out so he can stretch his legs. It doesn't take much. He likes to skateboard.

ME: *Does he?*

ADAM: Apparently so. I see him going up and
            down the hill. Then at six o'clock in the
            evening on the dot, he bounces back
            into the wardrobe. He likes it there.
            It's cozy. He's made a little nest in my
            T-shirts.

ME: *No wonder your T-shirts smell like that.*

ADAM: (calling out toward the closet)
            I love you, midget!
            Next year I'll give you a name.

Other times, STM's utterances inspire Adam
and me to examine life's important questions:

> **"Jesus nipples on ice! I am NOT going shopping for hamster wigs!"**

ME: *Are hamster wigs wigs for hamsters, or wigs for people made out of hamster fur?*

ADAM: Ooh, that's a good question. Well, if it was the latter, how many hamsters would have to be used?

ME: *Well, is it a toupee, or is it a long-hair wig?*

ADAM: It's a patch job.

ME: *Then maybe you'll need, like, six hamsters.*

ADAM: I reckon it sounds like I'm shopping for my hamster. My hamster needs a wig.

ME: *Is it for Halloween?*

ADAM: I was actually thinking it was for his self-esteem.

ME: *Awwww.*

ADAM: Little baldy hamster.

ME: *Why does he have low self-esteem?*

ADAM: He's bald!

ME: *Ok, now, a hamster wig, is it just for the hamster's head, or is it the whole body?*

ADAM: I don't know. I'm just imagining this tiny little hamster with an ill-fitting, wrong-colored little head wig. But he's happy.

After this conversation, we Googled "hamster wig" and, amazingly, came across a picture exactly as Adam had described.

Some people just have too much time on their hands (. . . says the woman who spends her free time transcribing recordings of her husband talking in his sleep).

Hold me.
I want you to feel greatness.

❦

I'm sorry, but, you can take your
can-do attitude and fuck it 'till it's raw.
Can you do that? Can you?

❦

Oh! It's a poltergoat. A poltergoat!
You can't see 'em, but you find all your
clothes chewed. If you listen carefully, you
may hear a ghostly baaaahhhhh.
Poltergoat! Baaaahhhhh.

❦

You must be a cunt. Or a lawyer.
Yeah, a lawyer.

"

CAKE-A-DOODLE-DOOoo!
It's cake for breakfast!

❦

She's knitting me a jumper.
Fuck! I don't want to be a social outcast.
Oh, not good.

❦

Ghosts going bump in the night.
Clumsy fuckers.

❦

Lead me not to the telephone, but deliver
me some e-mail.

"

" I'm like a vulnerable fawn in the woods. One that happens to carry an uzi, some ninja throwing stars, and a motherfucking bazooka.

♥

I'm totally too bad-ass for tango.
Cha cha cha!

♥

The carrots are winning! Damn those parsnips and their stupid infighting. They've got so much to learn. Bring on the swede. Ooooh, that'll show 'em.

♥

Oh, don't worry, dear. The spot doesn't make you ugly. No no no. The rest of your face, now THAT makes you ugly. The spot's just a highlight. **"**

**Ha ha ha. Who's crying now?
No, not you, you've got no tear ducts,
you tearless freak!**

❦

If you look at me again, I'm gonna bugger
your fucking eyeballs and eat them,
so you stare at my shit.

❦

Flowers for the lovers, schnapps for the
thinkers, death to the vegetarians ... I don't
care if you eat fish, you give yourself that
stupid name, you deserve all you get.

❦

What do you think you're doing? Totally
inappropriate behavior. Now sit down, put
the electric sander away and concentrate
on your art project, please. . . . Monkeys!

> **Cuff him! Arrest him! I don't care, that manatee is going down!**
>
> ●
>
> What do I think? Oh, I think as soon as I finish this sentence, I'm gonna kill you.
>
> ●
>
> I want Viking horns. Fuck-off big pointy ones. Yeah, Vikings. I've got an urge to pillage your ass.
>
> ●
>
> Don't. Don't! Oh, don't exfoliate your labia.
>
> ●
>
> **Everyday I wake up and I think, I look more and more like the perfect me.**

"

Fucksticks! That's it, I'm not playing anymore. Just give me back my tiara and my sash and the purple monkey. I'm outta here.

❦

You're never too old for Legos. . . .
Suck my balls, dumbfuck.
Building shit is fun.

❦

If you make me read Plato, I'm gonna punch you in the penis.

❦

Look at them staring at people like that. Your boobs are so obnoxious. Uch.

"

"

I've weaponized this pumpkin. Yeah.
Just for you.

Never before have I had the opportunity
to ride one of these wonderful creatures.
I'm gonna take it slow, and make
it last all day. Mmmm-hmmm.
I love saddling up my hamster.

Oh, stop crying, emo. You can write it all
down if you want. Then at least I don't
have to listen to your fucking whining.

Oh! It's so cute.
Now put it back in the fucking box.
It makes me want to puke.

"

"

Oh for fuck's sake! Double chocolate-chip cookie doesn't mean two chocolate chips per cookie! You're so literal! God!

♥

I'm so lucky to have disciples like you— FRIENDS, friends like you.

♥

Okay, Jesus, if you are the son of God, wave your hands in the air . . . Ha ha. Didn't think so.

♥

You're a cock and a fuck-up. Any further complaints can be directed toward my ass, where I'm sure you'll receive a warm response.

"

"

Yeah I want a bike with 128 gears.
Fuck off, I'm not gonna ride it, schmuck.
I wanna BRAG about it.

"

> "You can stop clapping now if you want. Really. You'll need your energy for cheering me later."

No question about it, Adam's alter ego loves an audience. I think Sleep Talkin' Man's taste for the spotlight goes right back to that February night when he gave his debut performance. I loved it so much, it's no wonder he stuck around. The more I delighted in his antics, the more prolific he became. Like a child, STM seems to thrive on positive reinforcement, and sulk when he's angry. On those rare nights when Adam and I go to sleep upset with each other, I never get a peep out of him, as if he's punishing me. On happier nights, he regales me with extended bouts of hilarity.

> "That's it. Your family:
> one big giant cluster-fuck."

Perhaps a look at the extended family in which Adam grew up offers some insight into STM's love of the limelight. This sizeable ensemble of uncles, aunts, cousins, second cousins, and second-cousins-once-removed frequently congregates, generally over food. These are high-spirited, raucous affairs at which the stereotypical understated Englishman would get lost in the overlapping cacophony of playful political debate, embarrassing anecdotes, and low-brow humor. But everyone in this crowd manages to hold their own. There is something very, well, Jewish about these lively family gatherings, which I found instinctively familiar and comforting, being so far from my own family and culture. It was heartening to find that, even in a society that is known for its reserved disposition, Jewish exuberance shines through.

Maybe it was a result of growing up as a member of this boisterous bunch that Adam developed both his joy in performing for others—something very much encouraged among the Lennard clan—and also his ability to comfortably fly under the radar when things get exceptionally lively—a useful talent amid such a rowdy group. As the sole performer in Adam's slumbering subconscious, however, STM can shamelessly hog the limelight all to himself.

I used to wonder why Adam's sleep talking so often coincided with my bouts of insomnia. I eventually figured out that it was precisely because Adam subconsciously sensed that I was listening—I might be tossing restlessly, or getting up to pee, or popping open the laptop—that he started performing. And, in fact, there is loads of evidence that Adam is on some level aware of the world around him when he's asleep, as he frequently reacts to subtle external stimuli. For example, this is how he responded to the sound of me typing away on the laptop:

"Clip clop clip clop clip clop clip clop clip clop . . . Who brought a horse in the bedroom? Oh well. Looks like I'm sleeping in the barn."

and me peeing in our en suite bathroom:

"What are you pouring that away for? I wanted that. Christ you're a selfish fucking cunt . . . I don't care. Cold tea is still tea nonetheless."

(um . . . yuck?)

and our little beagle Molly chewing her foot:

"Nibbling. Stop your nibbling.
Always with the nibbling.
Nibble my fucking fist! That'll stop your
nibbling. Can't nibble with no teeth."

and me getting back into bed:

"Whoop! Elephants landing!"

(I especially appreciated that one.)

He's even reacted to several stages of stimuli, like so:

> (I open the laptop in bed)
> "Mmmm, balls of light. Bibble bobbly bibble. Bibble bobble." (I quickly close it again out of guilt) "Oh, it's burst! I hope it enjoyed itself whilst it was around."
> (I think, screw the guilt, I'm bored, and open the laptop again. Adam sings:) "Here comes the sun. Hmm hmm hmm hmm. Here comes the sun, and I say . . . Hmmm. I said something. I forget what. Not important now."

My appetite for Adam's sleep talking became insatiable, so naturally I started exploring whether I could trigger him. After much experimentation, I have learned the following:

- DON'T say his name, or actual words of any kind. Something about language from outside his own head seems to jolt him back into the waking world, with all hope for sleep talking dissolving into the night.

- DON'T pet him, stroke him, kiss him, hug him, blow on him, or balance objects on his head. Any of these actions causes Adam to believe himself to be under attack, resulting in a violent awakening accompanied by the shouting out of the name of whatever creature is out to get him this time.

- DON'T wiggle around to shake the bed. Although this does get him talking, he whines about being seasick on a boat, thereby making me feel like a bad wife.

- DON'T pick up Molly the little beagle and snuggle her up against him. He is likely to start awake suddenly thinking she is a furry monster, and scare the shit out of her. I eventually concluded that the safest yet most effective way to get Adam talking is to make a short little noise that I somewhere between a grunt and a hum. Just to let him know, "I'm here. You can entertain me now." I call it "grumming."

I carried on with grumming for a while, with great results. Eventually, though, Adam noticed my grums in the recordings and figured out what I had been doing. Feeling like a guinea pig, he extracted a promise out of me that I'd stop. Being the loyal, obedient wife I am, I did stop . . . for a while. But sometimes, on those really quiet early mornings, when the light is just starting to peek through the window and I'm dying for the kind of comic relief that only Sleep Talkin' Man can provide, I give a quiet, little "maybe he won't hear this on the recording tomorrow" grum. It still works.

"

Yeah, Happy Valentine's Day.
Thanks for fucking me.

•

Now I'm going to ask really nicely
for you to un-fuck this situation.

•

Just shut up.
Your voice is drowning me
in a wave of bullshit.

•

Yesterday I made history.
Tomorrow can suck today's dick
as far as I'm concerned.

"

"

Guinea pigs make terrific drinking buddies. Just don't eat them!

●

I bring sassiness, sexiness, and awesomeness to the table. What the fuck do you bring, dickhead?

●

I really wanna kiss your face . . . with my fucking fist.

●

Steady . . . steady . . . focus . . . FUCK!
Concentrate, concent—oh shit! SHIT!
Hmm, there we go . . . Ther—
Oh ass cunt wobbly tits! I give up.
I'll never balance this penny.

"

"

Yeah, shove it in sideways
until you can't see any sunlight.
It may hurt, but it's the only way you're
gonna keep those potatoes dry.

●

Who put the broccoli with the
papaya?! Don't EVER put the
broccoli with the papaya.
The papaya needs no friends.
Leave the broccoli out of it.

●

Oh, I love the space you leave behind when
you go away. So please, fuck off and give
me back that space.

●

My eyeballs are sticky. Will you lick them
for me? Make sure you get all the way
into the tear ducts. Mmmm.

"

"

You over there,
you're going to be my friend today . . .
I don't know about tomorrow,
let's just get through today, asshole.

❦

Yeah. I'm gonna cook monkey brain.
Tell them it's cauliflower.
I love the sound of veggies
retching in the toilet.

❦

Don't you give me those puppy eyes.
Put 'em back. Puppy needs them.

❦

I am the hummus of knowledge.
And you are the breadstick to dip.

"

"

Bring it on, King Kong. I'll kick your monkey ass right back to the jungle.

♦

Little hands can't manhandle. No. They can only minihandle. Oh, pity those little digits.

♦

My ass rocks. Don't you just love it? You love my ass, can't take your eyes off it.

♦

I've got a horrible urge to catch tuna in your stockings. Sustainably, of course.

"

"

I am the lord of all pirates! I've got the
treasure map to find ALL treasure maps.
Beat that, suckers!

❦

You try so hard, and you fall so
short. Just give up, numbnuts.
It will save you a lifetime of pain
and shame. Your life: fail.

❦

Why don't you come back to me
when your brain's decided that it wasn't
designed to be a shit box, okay?
Run along now.

❦

It's NOT a fat ass.
I suffer from Hippo-bottom-mass.
You should feel sorry for me.

"

"

You know, it's not easy being me.
You should try it. I bet that after just
five minutes, you'll then have
an incredibly healthy respect for
how amazing I am.

♥

OK, so that's your weekend homework.
Go home and slap grandma.

♥

Come to me my penguin brothers
and penguin sisters
and dance your dark winter dance
and flap
and flap
those fin-like wings

"

Oompa loompas don't sing in heaven.
They tidy up the clouds.

❦

Nobody told me I couldn't bring that
to tea. Why can't I bring it to tea?
I'm really sorry to have upset you.
It's not my fault, I thought it was OK.
Oh, it's not fair. You're bastards
and you're trying to spoil my fun.
Go on. I'm taking the iguana home.

❦

**Come on. Let's all go be happy in
front of some miserable people.**

"Where—Where's my coconut?
I know you moved it. Give it back to me.
I want my coconut! I can't pee without my
coconut. I'm going to peeeeee. Oh,
I'll have to get a new coconut. Mmm, that
one was just the right size . . . and furry.
Coconuuuuuut . . . coconuuuut . . . where
aaaaaarre yooouuu? Mmmm. Bollocks."

ME: *You said, "Where's my coconut? I*
       can't pee without my coconut! It was
       just the right size. Coconuuut . . .
       coconuuut." You just went on and on.

ADAM: Well, they may all look the same, I'm
         sure coconuts feel they're individual.

ME: *If coconuts feel enough to know they're individual, then how do you think they feel about being used as a receptacle for urine?*

ADAM: Whoa, whoa, whoa, you never said I said that. You said I can't pee without my coconut. I may need it as a security blanket. And have it next to me when I pee. And its three eyes can be looking up at me saying, "Good boy."

> "I like the balloons. I want one.
> If I don't get one, I'm gonna squeeze
> one out right here, right now!"

Earlier, I compared my sensitivity to Adam's sleep talking to that of a mother to her child's cry—and it now occurs to me that STM is, in fact, very much like a child. A precocious, obnoxious, extremely foul-mouthed child. Consider his complete lack of inhibition, irrepressible playfulness, disregard for consequences, random zaniness, capacity for cruelty, and yes, those sulky tantrums.

We sometimes find STM in typical childhood situations familiar to many of us:

> "Don't pick me last. No, please.
> Oh, you bastard! Now I'm not going
> to play ... I know I'm sulking.
> You can fucking suck my fat one."

(Although I don't know how many kids could manage to face such a crushing ego blow with that combination of both self-awareness and bold retaliation.)

Other times, STM is happily lost in a world of fanciful imaginary creatures the likes of which would impress even Dr. Seuss:

"No, I want to swim with the giant gajumba. Hold on to their shell . . . the ones with the spiky faces, you idiot! They're fun."

Or we may happen upon him participating in one of those fiendish activities sadly common among the less compassionate of young boys:

"Catch it. Catch it . . . use both hands . . . its legs are still wiggly! Now clap. Niiiiice."

STM gets to act on every childish impulse, without any niggling parents or pesky adult conscience trampling on his fun. Freud might call him the ultimate id.

STM emulates a very different sort of child than Adam was himself. And what was Adam like as a kid, you ask? Let me set the scene.

> "Welcome to your first day at duck school. I'll make the lesson simple.
> OK: Quack. Quack quack. Very good class. Now go swimming."

England in the 1970s. There's an alternative notion of teaching afoot. The idea is that children learn best through play. Many schools had adopted this educational theory with promising results. Adam's first school took this premise further than Maria Montessori ever intended. It seems their approach could be summarized as "children should just play all the time; they'll learn stuff eventually, right?" Riiiiight.

Adam, a joyful, creative little soul, had taken the "play" aspect of this philosophy and run with it, and had consequently become a darling of the school. Throughout his early educational career, his parents received glowing (if vague) reports regarding his progress and talents. They were so pleased at how their younger son was excelling academically. At the age of six, Adam took the entrance tests for the new school to which his parents hoped to send him. Faced with this novel situation, Adam made his best guess, based on previous experience, as to what was expected of him. You can imagine the dismay of the test marker to receive an exam back that was covered in intricate, lovingly drawn doodles, but didn't offer an actual answer to one single question. Well, what would you expect from a child that had never taken a test in his life, and—by the way—couldn't remotely read or write? The letter the prospective new school sent to Adam's parents suggested that perhaps their son was retarded. Adam did get accepted to the school in the end, after being sent off for a full battery of tests with a psychologist to prove that he was not, in fact, mentally challenged. Here is his father's response to the acceptance letter, which I came across while going through Adam's old school records:

Dear Headmaster,

Thank you for your letter of the 2nd March with the good news that Adam has won a place at your school. I enclose herewith duly completed the acceptance form with the acceptance deposit. I trust you will excuse the teeth marks in the acceptance form. The family's general air of excitement transmitted itself to our dog, who managed to get to the post with the ensuing results.

Yours sincerely,
R. H. Lennard

One wonders if perhaps the headmaster thought the whole family was a bit special.

**It's the belly button button.
Press it and go to hell!**

> "Of course I've always loved music. YOURS makes me want to poo my pants."

As Adam entered the double digits, the free-spiritedness that defined him in his younger years started to give way as the insecurities of adolescence crept in, and the little embarrassments and setbacks of secondary school did nothing to help. When it was time for Adam to begin secondary school, he joined his older brother Darren at Haberdashers' Aske's School for Boys, a posh private school with a strong focus on academics (play having been proven inadequate preparation for our non-play-centered world). It may not have been quite the right fit for sensitive, creative Adam, but he made an earnest effort to find his place.

One such endeavor was to join the music department. Adam had always been interested in percussion—in fact, he would later go on to become a downright decent drummer. But these were very early days in Adam's music education, and his teacher thought it best to start him off on the forgiving triangle. This instrument propelled Adam into an elite society of the school: the Haberdashers' Aske's third orchestra.

(Note the "third." That means there were two that were better. To get a sense of the level of talent—and enthusiasm—we're talking about here, go to YouTube and search for "Orchestra Fail." Make sure your speakers are turned up really loud.)

Adam's parents remember with great relish Adam's first (and last) concert with the orchestra. Seeing him up there on that stage, surrounded by violinists, cellists, and flutists, the Lennards were bursting with pride. As the conductor tapped his wand on the stand, Adam raised his triangle with an air of gravity, preparing himself for his triumphant debut.

It just so happens that, in the piece being performed that evening, the triangle was to remain silent for the first 148 bars, only entering in the final climax. For 148 breath-holding bars, Adam's parents watched him, arm held high in the air in front of him, clutching his triangle aloft—looking for all the world like one of the Hitler Youth—nodding his head dramatically to each beat as he carefully counted the bars. Finally his moment approached, and it was with an expression of triumphant ecstasy that Adam raised the striker, and struck that triangle with all the passion he could summon.

Unfortunately, Adam had forgotten that he needed to hold the strap on which the triangle is hung, rather than the triangle itself. So, when he finally struck his fair instrument, he produced nothing more than a dull, discordant clang. From his place on the stage, Adam saw the front section of the audience recoil in horror at the cruel assault on their unsuspecting ears.

Thus came to a close Adam's tenure with the Haberdashers' Aske's third orchestra.

As he got older, Adam, still looking to find his place, took a stab at mixing in with the "bad" kids. One day the school board members had come for a meeting, and Adam's friends decided it would be amusing to steal all of their Mercedes hood ornaments. When the theft was discovered, the headmaster rounded up all of the usual suspects, Adam among them. The first kid emptied his bag—out clattered a Mercedes ornament. *Clinkety clank.* The next kid, same thing. *Clink clank clinkety clank.* One after another, all of the boys were forced to dump out their prized booty on the headmaster's desk. Finally, the headmaster faced Adam at the end of the line. Adam sheepishly upended his bag, and out clattered . . . a Ford ornament. *CLUNK.* He had been so worried about the damage he would cause to an expensive car, he had passed up all the shining Mercedes for an unassuming Fiesta. Thus ended Adam's stint among the hoodlums of Haberdashers', such as they were.

From all family and school reports, Adam was a delightful little boy: sweet, funny, harmlessly mischievous, creative, outgoing. But alongside those traits, he was also a sensitive, emotional child, and as he got into his formative years, that sensitivity transformed into a consistent struggle with low self-esteem. Recently, I got a chance to go through all of Adam's old school records. I expected to find lots of humorous comments from teachers proving just how much of a character he was. Instead, what I found over and over, year after year, was some version of "Adam is a wonderful, bright boy with lots of potential. But he needs to work on his self-confidence."

Now, many soul-searching, therapy-attending years later, Adam has a very healthy sense of self. He's secure in who he is and genuinely happy. But what does this all have to do with the sleep talking? Well, it turns out that many psychologists believe that, in the deepest layers of the mind, lots of the growing up and maturing that we all do doesn't penetrate. So, somewhere in his subconscious must live child Adam. And perhaps, when adult Adam is asleep, when his conscious mind lets down its barriers, the sensitive, insecure Adam of his youth is left exposed. But now, in a playground all his own, he has the opportunity to do it all over again, as a guy who has the balls to stand up to anyone, who says what he wants, swipes whatever metaphorical hood ornaments he wants, who knows that he's the best damn thing that ever walked the face of the earth. And you know what? If that helps my husband claim back some of the tougher moments of his childhood, I'm all for it. Because in those wee hours of the night, when STM appears, that triangle rings out as clear as day.

STM: MASTER OF SELF-AFFIRMATION

10 "I'm the reason why there's so many adjectives for awesome."

9 "Well, let's face it: I'm so good looking, even my bacteria are cute."

8 "There's only one thing that comes close to being as fantastic as me, and that's my reflection."

7 "Yeah, keep looking. It doesn't get any better than this."

6 "Here I am! Perfection on two legs."

5 "I'm not just good. I'm lick-my-parts tingling kind of good."

4 "It's not blood that courses through my veins. It's incrediblood!"

3 "I'm better than Superman. He's just a cunt . . . in underpants."

2 "I'm so magic, I puke rainbows and shit pixie dust."

1 "If beauty is in the eye of the beholder, then BEHOLD, BITCHES!"

66

There was so much blood!
Oh, there must have been at least
five llamas. Totally unprovoked attack
by those puffins. I managed to clip their
wings. This is llama turf.

❧

Not being able to do something could teach
you a lot about yourself. Mmm-hmm.
Like what a fucking loser you are.

❧

Really? If you can pee that high,
DEFINITELY join the fire brigade. Yah.

❧

Darling, with an ass as big as yours,
innocent bystanders could get hurt!

99

"

I am Mediterranean Man! Hear my cry:
TZATZIKIIIIIIiiiiiii and tremble with fear.
I will cut you with my throwing pita.
But not at three o'clock, 'cause it's siesta
time. TZATZIKIIIIIIiiiiiii and away!

❦

Now I'm going to batter you to
death with chicken drumsticks!
It'll be really messy, but I'm going
to enjoy every hour of it.

❦

Tiptoeing elephants? Come on guys,
give it up. I can see you! You're huge!

❦

Stand further away. You can't possibly
appreciate my greatness this close up.

"

"

I'm like a god. Only, it'll hurt more when I judge you.

♥

Listen: Some people play Scrabble.
Some people play chess.
You? You play turd puppets.

♥

Well that's just great. Peanut butter in my crack. Goddamn it.

♥

I've written your epitaph. Yup.
I did it early. You wanna read it?
"Here you are, lying dead. Ha ha ha ha ha."

"

Hey, boobs! Stop staring at my face!

,

Oh yes, I must have an enema.
And I'm going to keep what comes out,
'cause it reminds me of you. I will take it
home, dry it, make it into paper, and write
your name on it as many times as possible,
and frame it. Put it on my wall, and there
it will remain. It will be my memory of you.
I may do it more than once,
depending on what I've eaten.

,

I need this like I need a second crucifixion.

It's growling. Shhh, it's growling closer . . . It's an angry thing, a big angry thing. It likes cabbage, though.

🖈

I'm not waving at you. I'm just building up for the big fucking slap you're gonna get.

🖈

I could go find somebody who could surgically remove that stick from up your ass. Or, you could just chill the fuck out, Batman. Choose.

🖈

Hey! You killed my velociraptor, dickhead. That's so unfair. You do realize how hard it is to find one of those 'round here, don't you?

"

You're a complete waste of space.
Just go home and apologize to
your mother's vagina.

•

I need a big room, with strobe lights.
And people riding bicycles . . . naked.
To classical music, of course.

•

Two hats for my bunny, please.
Make 'em smart ones.
He's got to look the business.
And no more fucking spats, OK?

•

Loving you is an important life lesson.
You learn about all the fucking
stupid mistakes you make.

"

"

Leave the broccoli alone.
It can sort out its own problems.
Confusion is part and parcel of its life.

♥

. . . Then out of nowhere, the puffin ninja
kicked my ass! Little fucking runt bastard.

♥

Where are we going?
I want to know where you're taking me.
It's all fun not knowing, but now I'm bored,
so FUCKING TELL ME WHERE
ARE WE GOING! . . . Ooh, I've never
been there before! I hope it's good.

♥

I'd say welcome to the School of Life,
but you wouldn't pass the entrance exam.
Dickhead.

"

"

All I want out of life is ice cream and cuddles.
Is it too much to ask? Is it?

'

Half the time listening to you, I'm imagining
the carnage of pulling out your tongue and
wrapping it around your throat.

'

Your singing can wake the dead.
So shut the fuck up. I don't want
any zombies dropping their jazz
hands all over the fucking place.
Alright? Just shut it.

"

Stupid-fucking-cunty-bollocks-expialidocious

> **"Yeah, falling in love is WONDERFUL. Especially when it's with me."**

By now you might be wondering how Adam and I met—especially with an ocean between us. I think it's actually a pretty juicy story. And, of course, it involves sleeping.

It was 1991. The Western world was in the early stages of recovery from the cultural atrocities of the 80s. Synthesized pop ditties, rock power ballads, and neon nylon had given way to grunge, flannel, and apathy. But in the Jerusalem nightclubs, it was Duran Duran and "Land Down Under" every night of the week.

I was spending a year between high school and university in Israel on a program with a Zionist youth group. Our year was split between studying in Jerusalem, teaching in a small-town school, and working the fields on a kibbutz. In those first months in Jerusalem, I spent the days soaking up the history, architecture, culture, and

107

language, and the nights dancing until the sun came up. Sleep was not on the syllabus.

Adam was on a similar program with a sister youth group from the United Kingdom. Occasionally, the leaders of our two organizations threw us all together for social weekends. You can imagine the bedlam: a bunch of teenagers out from under their parents for the first time, in a country with no legal drinking age, crammed into a dorm with another bunch of teenagers with exotic, and therefore inherently sexy, accents. It was hormonal pandemonium.

Aside from treating each other like foreign cuts of meat, I made some good friends among the British guys at those international gatherings, and it wasn't long before I was spending lots of my evenings in their dorm rather than mine, watching them play Risk for hours on end. (Wait, did I say hormonal pandemonium? Maybe I overestimated British teenage guys' idea of a good time.) That's when I first noticed Adam.

Adam has a distinct memory of overhearing me whisper into a friend's ear, "My God, check out his lips," and knowing that I was talking about him. Yup, he was right. Even then, Adam had these gorgeous, full lips that just cried out to be gnawed on . . . But I digress.

It was a cool evening when we all set out from the Brits' dorm to walk to the dance club in Talpiot. Until then, Adam and I had only admired each other from afar. But he had decided that, on this walk, he was going to speak to me. And so he strode up alongside me, and we started talking. To this day, although neither of us can recall a word that was said, we both remember with crystal clarity how immediate the connection was, and the breathtaking excitement we felt. We talked and talked, until we hit the club and the music drowned out all conversation. So then we danced and danced, rapturously, illuminated in the blues and reds and stark white of the ever-shifting club lights, unaware of anyone or anything else. It was electrifying.

We walked home together at six in the morning and stopped at the beautiful Talpiot overlook to watch Jerusalem turn gold with the sunrise. We both knew that something extraordinary was brewing.

From that night on, we saw each other when we could, and got to know each other in bits and pieces. Meanwhile, I had parted from my program and found myself an independent kibbutz to finish out my year. So when Adam's group was coincidentally sent to my kibbutz for the weeklong wrap up of their program, it seemed that fate had intervened on our behalf. I was nearly paralyzed by anticipation and nervousness waiting for him to arrive. Had everything we'd experienced in the whirl and excitement of our brief episodes together been real? Could it be sustained beyond those moments on the dance floor or watching the sunrise?

The day arrived. I didn't know exactly when Adam would be arriving on my kibbutz that evening, so I went for a late-night swim (read: scaled the fence around the pool for a skinny-dip) with my friends. When I returned, I opened the door to my room to the surprising sight of Adam sound asleep in my bed. I had never seen anything so entrancing. I couldn't explain what I was feeling, but it was colossal. I watched him sleep for a long time.

As Adam tells it, he arrived on my kibbutz a bundle of nerves, and realized that he had no idea how to find me. He asked around until he was directed to my room, and let himself in. There wasn't much else in the room besides the bed, and so he settled himself down for a bit of a rest while he waited. I find paraphrasing him way too embarrassing, so I'll let Adam speak for himself here: "I got into your bed, and suddenly, my whole body seemed to be crackling with electricity. I could smell you in the sheets and the pillow. I closed my eyes to breathe you in and soon, exhaustion overcame me. The next thing I knew, you were waking me up, all cool skin and wet hair and sparkling eyes. I had never seen a more beautiful sight in my life."

It was in that moment that we fell in love.

Adam blew off all of his group activities to spend every possible moment with me. I'd get up at four-thirty in the morning to go out to the fields, make it back by eleven a.m., and the rest of the day and evening was ours. We lost all track of time when we were together.

I remember so clearly the day that Adam's

group was leaving to go back to England. We had already kissed goodbye a thousand times, the bus had been boarded and was lumbering away down the dirt road while I wept and waved. When the bus stopped at the kibbutz gate to wait for someone to come out to open it, Adam threw open the emergency window at the back of the bus, climbed out, and ran back up the road to steal another kiss.

(Sometimes I amuse myself imagining how STM would have done it. I envision him leaping out of the back of the bus, sauntering over, looking me dead in the eyes and declaring (as he has in the dead of night), "Kissing's good for your health. So pucker up, baby, I'm gonna make you live past a hundred!" Not quite the stuff of a great romance.)

Adam and I soon found ourselves back in our respective countries, trying to navigate a relationship from opposite sides of the ocean. Remember, this was before the Internet; there was no e-mailing, IM'ing, text messaging, video conferencing. Transatlantic phone calls cost a fortune, especially for a couple of nineteen-year-olds. But it was also before the death of the written letter, and, even better, audio cassettes.

Yeah, remember those? Adam and I raised the art of the mixtape to new heights: our foremost form of communication, we mailed back and forth recordings of us talking, intermixed with music. I so clearly remember those desperate trips to the mailbox numerous times a day, hoping to hear from him, and then the exultation of opening a package and sliding the cassette into the stereo. Accustomed as most of us are now to anytime, anywhere communication, it's not often that we have the opportunity to experience that bittersweet, agonizing anticipation. Even back then, Adam was so emotionally expressive, so clearheaded in his feelings. I would listen to his tapes hundreds of times, until I had every word, every breath memorized.

I was prepared to do anything to be with Adam, even wait. But for him, it was too hard. As he explains, "I was crazy about you, the feelings were overwhelming, unlike anything I had ever experienced. But at nineteen, I couldn't bear the pain of being separated from you, of having those yearnings continually unfulfilled. I dealt with my emotions by burying them, until I could leave you behind. I buried a huge part of myself in the process." After a few months, we parted very painfully.

My heart was broken. A part of me harboured the hope that Adam would find his way back to me. Meanwhile, I dated my guts out, looking for someone to fill the void he'd left behind. But no one that I met could measure up to my memory of him. After two full years of anguished pining, I finally moved on and committed to a life without him.

Of course, being the master of timing that he is, it is precisely around that time that Adam came to his senses, pulled himself together, and decided that he had made a terrible mistake— that we belonged together. He began calling and writing, trying to woo me back. But for me, having worked so hard to get over him, it was too late. My heart was closed to him.

And thus, it was Adam's turn to hope and pine. For months, he waited every day for the phone to ring. And then, finally, he went on with his life as well.

TOP TEN
STM Pick-up Lines

10 "Kiss me. Tastes good, doesn't it. Why don't you go back and have a second helping? Be greedy."

9 "Keep close to me. My love is infectious, and there's no cure."

8 "Of course I know where your eyes are. I just like staring at your tits."

7 "Let's swap saliva. I know, love is messy."

6 "You're in the crosshairs of my love. And my aim is true. Bitch."

5 "Well, don't YOU look like the icing on a fuck cake."

4 "Nothing's invisible to my love radar. I'm picking you up loud and clear."

3 "Your three steps to happiness: Sex, Food, Me."

2 "Sure you can have my phone number. It's like having a direct line to God. But better. Because I answer."

1 "Rub my tummy, bitch."

2007, New York City. Twelve years had passed since we had last spoken, fifteen since we had seen each other. I had long ago moved on with my life, and had a number of serious relationships, none of which were quite right. I had thought of Adam occasionally, but only as a distant element of my past. And then, one day—on a momentary whim—I did what millions of ex-girlfriends have done the world over: plugged his name into Facebook. And there he was! Swiftly, without too much thought or analysis, I dropped him a friendly e-mail.

Adam's response was immediate and enthusiastic, and we charged into catch-up mode. He had gotten married (Oh.); he had gotten divorced (Oh?); he had two children (Ohhh. Well . . .). We graduated swiftly from e-mail to IM, to phone, to Skype. We spent hours each night talking on video, without any acknowledgment that anything romantic was going on. The official unspoken line: "Oh, this is normal, we are just old friends who happen to spend FOUR HOURS every night Skyping. Nope, nothing to see here." What a feat of mutual denial it took to pull THAT off.

One day, I suggested that we needed to get together for cup of coffee. Just so we could, y'know, settle the past and put it behind us. He agreed. So he did what any sane, totally-just-friends person would do under the circumstances: purchased a plane ticket, London to New York. The most expensive damn cup of coffee in the history of the beverage.

It was midnight, December first when I arrived at the airport to pick up Adam. Yeah, let's review that. He's coming for a CUP OF COFFEE, we've never acknowledged that there's this little matter of WHERE HE'S GOING TO SLEEP and his flight gets in at MIDNIGHT. So there I was, waiting for my totally platonic friend from London to drop in for a nice cup of coffee. The nerves were killing me.

Of course, Adam got held up in customs. It seems his insane-hyena-on-ecstasy grin caught the eye of the officers who probably thought he was high or crazy or both. They pulled him aside and started grilling him as to the nature of his trip, in response to which he told them our entire love story thus far. Adam must have successfully appealed to their romantic sensibilities, since those customs officials did eventually send him out to the airport lounge where I was waiting for him.

Adam says, "I had never been so hyped-up with anxiety in my life, charging through the airport, getting though customs. But then I came through those doors, and the moment I first saw you, I knew that I was in love with you, had never stopped loving you, and that I wanted to spend my life with you."

I, on the other hand, couldn't look at him. Literally. Even though I had been staring at his face in 2–D for hours every night, there was something so significant, so momentous, so goddamn BIG about seeing him for the first time after all those years, I just couldn't do it. We stood there toe-to-toe, my eyes on the linoleum, for twenty minutes. And the longer I waited to look up, the BIGGER it became, until we had passed beyond the realm of the dramatic, into the territory of the utterly absurd.

Finally, just to put myself out of my misery, I lifted my chin and flicked my eyes to his, and then as quickly as possible buried them again against his chest in a smothering hug. It was a lot easier driving, when I had an excuse to only glance at him for fleeting moments. By the time we arrived in my neighborhood, I could almost look him full in the face.

Six weeks later, we were engaged.

66

When are you gonna learn:
when transporting giraffes, shouting "duck"
every time you hit a bridge doesn't work!

99

Elephant trunks should be used for elephant things only. Nothing else.

'

Mmmm, I do like your shampoo.
Smells like llama spit. Is it llama spit?
Smells like llama spit.

'

Listen up, people.
Sasquatch has been in my sock drawer again,
and now he's somewhere lose in the bathroom.
So I'm setting up a six-foot perimeter fence,
and someone bring me my clippers.
He's mine this time.

Hey, I know you, but I don't like your face.
Take it off . . . That's much better,
much better.

●

You can't do that to a jellyfish. I'm calling
the authorities. You dirty jelly molester.

●

I'm just a chubby ninja.
Able to move between skinny people.
Tiptoeing elephant. No one can see me.
And then I attack! With ice cream
and jelly, with chocolate sprinkles on top.
Mmmmm.

●

How do blind people know they're
done wiping? How?

"

I've got to get me two or more of those little fellas if I'm going to be able to make goblin soup. Mmm-hmm. It's so true.

If that shark thinks it's coming to bed for a cuddle, it's got another think coming. God, it's all me me me with that fucker.

Drilling . . . drilling . . . keep on drilling . . . drill . . . Oops! Fuck me.

Now this little piggy went to market. And this little piggy had roast beef. No, this little piggy had none. Because this little piggy was vegan. He doesn't eat roast beef. He's weak. He's easy catching.

"

"

Has everybody got their gerbils? Alright?
Good. Okay, commence shaving!

●

**What the fuck's wrong with
your face?! Christ on a bike,
next time you're smiling, warn me.**

●

As incredible as me.
As spectacular as me.
As awesome as me.
Oh, the similes are just so limiting. Really.

●

Oh, this is a one-man job. A very big man
with six arms and enough ears for each one
of your fucking suggestions.

"

"

Where'd you put the pelican food? Where'd you put the pelican food?! They're gonna be really cross if we don't get the pelican food. Shit. I found the toys for the toucans, but I need the pelican food. If you don't find their food, I'll find a reason to punch you in the face. I'm not going out there without the pelican food. Fuck it, you can go out there. You explain it to them. They're all beaky and flappy. Once they get excited, they just cause a mess. There's no reasoning with them, either. Nope. Unreasonable beaky twats. Big mouth fuckers. Find their food!

●

We should stand up for the downtrodden and ugly. Let them taste freedom for a moment . . . and then nuke the cunts.

"

"

I can see your future.
And—oh. You just had the best bit.

♥

Talk once more, and I will sue
you for ear abuse. Shame on you.
Shame! Auraphile.

♥

The noodles are coming!
Stand by your chopsticks.
Steady the miso . . . Steady . . .
Don't do anything until you see
the whites of their strands.

♥

It'll suck to be a moose.
You just can't play Pong.

"

"

I'm sorry,
I can't come to the phone right now.
But if you're not my mother,
you can leave a message. Beeeep.

,

You didn't leave me any ice cream.
Why didn't you leave me any
ice cream? You shit-head mother
fuckers, you never give me any
ice cream. Well happy fucking
birthday. You're fucks, the lot of you.
I only wanted some ice cream.
With chocolate sauce. And Oreo.
And marshmallow. And some
chocolate sprinkles. Yeah, that's
all I wanted. AND A SPARKLER.
Couldn't even do that for me.
I hate you all.
FUCK BAGS!

"

"

Damn those tortoises.
They outnumber us fifty to one, and we're
not allowed to carry stun guns! Oh!

❦

Milkshake time! Everyone grab your cow!

❦

BLECH!
You can't give me that cottage cheese
shit. It's like albino diarrhea.

❦

No puppy! Bad puppy!
Make you into puppy slippers.

"

One of the fun unexpected side effects of STM going public is that we've heard from and met some of the most extreme sleep talkers and walkers in the world. Compared to some of them, Sleep Talkin' Man seems downright subdued! Once, on a talk show, we met a woman whose antics were truly impressive. On the mild end were the long bouts of sleepwalking where she'd pace angrily around the house wielding kitchen knives, or methodically wash all of her panties in the cat's water bowl. On the extreme end was a story she told about the night that she went into her grandmother's room, dragged granny out of bed, forced her into the bathroom, and tried to give her a bath.

So many people have written in to share their sleep talking and sleepwalking stories with us. It seems almost everyone has a tale about the bizarre things their bedmate, sibling, parent, or college roommate has done in the late-night hours, and, yowza, there are some shockers! An entirely unfettered subconscious can prompt some astonishing behavior, including breaking into your neighbor's and crawling into bed with them, and (ew!) drinking the oil off the top of a jar of natural peanut butter.

I've sprinkled our favorite fan stories that we've received throughout this book as "Letters to Sleep Talkin' Man." I love this stuff. To me, the best thing about these anecdotes and the others we've heard is that (usually) no one is hurt by anyone's sleeping antics and, in the end, everyone's got a great story to share.

Letter to Sleep Talkin' Man

One night I was having this dream, but it was the kind of dream where it feels like you're actually awake. So as far as I knew I was awake, but I was actually sleepwalking. I went into the bathroom and saw the toothpaste lying on the counter, missing its cap. I don't know why but for some reason this was the WORST possible thing that could EVER happen and if I didn't find the cap IMMEDIATELY, the world was going to END!! So I looked everywhere—on the floor, under the cabinet, everywhere. No cap. Time was running out! Then I saw the roll of toilet paper and had an idea . . .

I grabbed the toilet paper and started wrapping it around the toothpaste. I wound it around and around till the whole roll was gone. So now I had a big ball of toilet paper holding the toothpaste in the tube, but NO! That wasn't good enough! So I went into my room and opened my dresser drawer and grabbed a couple T-shirts and wrapped those around, too. Finally, when I had a bundle the size of a football wrapped around the toothpaste I felt relief. The world was finally safe! I put the giant toothpaste tube–ball into my drawer, closed it, and went back to sleep in peace.

In the morning I woke up to the sound of my dad yelling, "WHEEERE'S THE TOOTHPASTE?" Suddenly I remembered the whole thing. "No, it can't be," I was thinking, "it had to be a dream . . ." But I opened my dresser drawer and . . . yep, there was the ball. So I started unwraaaapping it and unwraaaapping it—it took FOREVER. When I finally got all the stuff off, there in the middle was the toothpaste . . . with the cap on.

Tammy S.
Ontario, Canada

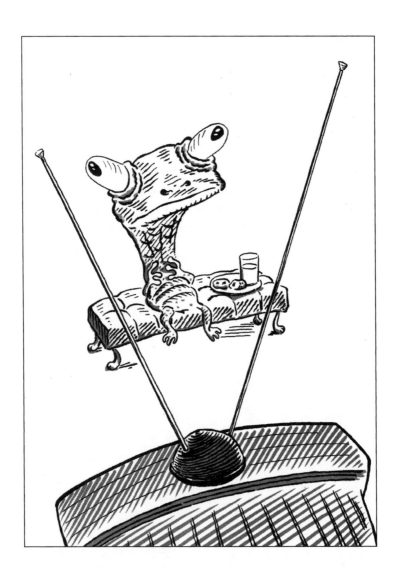

"

Smug fucking chameleon,
with its googly oogly eyes.
Stick it in front of the TV.
That'll fuck it up.

"

> "Going to husband school. Always having to do better. The teacher's a bitch and there's no chance of graduating."

I can't tell you how many concerned e-mails and comments I've gotten from blog readers, checking to see if I'm emotionally intact after the latest STM zinger. For example, the above. That's right, lying right next to me, Adam said that in his sleep. But no, he did not wake up to find me sobbing dejectedly or glaring accusingly. Granted, Adam does say some truly dreadful things in his sleep. And I suppose that there are some women out there who would worry that their husband was revealing some dark, hidden feelings about her that he dare not utter in the light of day.

I am not that woman. There has never been a single utterance to come out of Sleep Talkin' Man that has offended me, or left me fretting for my relationship. I've got a husband who still sends me off to work with sandwiches in which he has tucked little love notes. Why would I get myself worked up over some nutty nighttime nattering?

STM does spend an inordinate share of his limited airtime spewing insults. People have commented that he must really dislike people, or suffer from some serious anger issues. All I can say to that is, as the person who sees more of him awake and asleep than anyone, it's just not true—Adam spends most of his days in good spirits, and genuinely likes and gets along with just about everyone. So I believe that, for the most part, his clever but scathing insults are not inspired by anyone, but are just Adam's subconscious letting off a bit of steam while entertaining itself along the way.

Nevertheless, Adam's sleep talking has occasionally gotten him into a spot of trouble with those who did take something personally. For example, this little time bomb from two years into our relationship:

> "Your mum's at the door again.
> Bury me. Bury me deep."

My mom read the blog that day. When she spoke to me a day or two later, she joked about Adam owing her an apology. By the time she had made this quip three more times over the next week, I got the message that she really was waiting for an apology, or at least some assurance. I think my mother was feeling genuinely concerned that STM was revealing some kernel of truth about Adam's feelings. Seriously, Mom, the guy herds pygmy goats, keeps a midget in his closet, and married a zombie. How much more evidence would it take to convince you that STM's universe is populated with an entirely different cast of characters than Adam's? There's a whole other world up there in Adam's head, and STM insults his way through every bit of it.

That said, it often happens that particular events in Adam's waking life crop up in some warped form in his late-night mumblings. This has been known to drop him in a bit of hot water. For instance, one night after he had shamelessly teased the receptionist at work about the prolific :) and :(faces studding her e-mails, STM came out with:

> "If she sends me one more fucking smiley face emoticon, I'm gonna shove that keyboard so far up her ass, she's gonna have to tweak her nipples to force quit."

That one caused some trouble. To be specific, when he walked into work the next morning, it caused a stapler to be hurled at his head.

Real life references didn't always pepper Adam's sleep talking as they do now. In fact, for a long time, everything Adam said seemed utterly random and unrelated to reality. It was a full ten months into STM's existence before he said anything that we were able to recognize from our waking life. But there was no missing it that first time:

> "My snorkel! My snorkel! Judge won't like it if I don't have my snorkel! You PLANT it . . . Don't want baby snorkels."

Now, I can't tell you where the snorkels fit in. I can tell you, though, that this was the night before he was going into family court concerning visitation with his kids. Perhaps his anxiety was so intense that it broke right through more literally than anything else had up to that point. Although, as you can see, he didn't quite manage a complete departure from the surreal.

We had another month of nothing but randomness from STM, and then, just after one of our best friends revealed that she was pregnant for the first time, we got our next little glimpse of Adam's life in STM's exposition:

> "Babies don't bounce. They don't bounce!
> Shame.
> It'd be much more fun if they bounced."

This regretful warning was most informative for our pregnant friend, I'm sure.

Another month went by with no midnight mentions of Adam's waking life. And then we went on our honeymoon, and Adam's two worlds collided! He babbled nearly every night about the day's happenings. Our honeymoon was a bit unusual, you see—we spent two weeks volunteering at an elephant sanctuary in Thailand—and thus, I believe, it provided plenty of fodder for STM's interest in the abnormal. It started with this one:

> "Oh, such wrinkly skin. And oh so hairy. Yeah, like grandmothers with trunks."

No real mystery what he's talking about there, although it does beg the question of how hairy Adam's grandmother was. Another night, we had this:

> "Stop bouncing the floor. Stop it, seriously, I need to pee. I need to pee and I can't pee in the toilet when you're bouncing the floor ... Fuck you shit-for-brains, that's it, I'm gonna piss up and down your body every time you bounce ... There we go."

This one was definitely inspired by real life: at the sanctuary, we lived in a raised structure made of mostly bamboo. When anyone from any of the huts walked around, the entire place shook with their footfall. This next one is pretty self-explanatory, given the fact that we were in Thailand:

> "What goes in one hole hot comes out the other hole hot. Burning fucking curry. Awesome stuff."

And finally, for anyone who has ever shoveled elephant poop, this one needs no explanation:

> "Totally green snowballs. Giant ones! They look wrong. They sound wrong when they hit you. And boy do they smell wrong."

I can only hypothesize that our daily experiences started appearing in Adam's sleep talking because in an elephant sanctuary, a situation as

unlike our typical life as you could ask, our day-
to-day was suddenly just as interesting as his
imagination. Our stint at the sanctuary seemed
to throw wide open the portal between Adam's
waking and sleeping lives. Even after we'd
returned home to "regular" life, reality cropped
up more and more often among the usual
bizarre, fanciful musings of STM. STM was
inspired by people Adam met, as in this utterance after an evening out entertaining a client
with an especially noteworthy beard:

> "My beard can tell a thousand stories.
> My mouth, however, just says, 'fuck you,
> stop staring at my beard, weirdo.'"

He was stirred by television shows we had
watched, especially our most guilty-pleasure
programs, like this little number during our
Ghost Hunters phase:

"I need someone else to help me catch ghosts. 'Cause we're going out to kick seven shades of spiritual shit. Yeaaaah. Ghost Kickers! Free floating vapor? Free floating fucker, more like. Come on, let's get 'em!"

And sometimes, he simply replayed—with his own twist—experiences Adam had, like this narration while we were spending a long weekend in Belgium, which is justly famous for its waffle houses:

"This waffle is my waffle. I will put my ice cream, my sauce, my cream, and even possibly my strawberries on my waffle. And you will watch me eat my waffle. You will watch me as I carefully slice it up bit by bit. You will watch me wipe my mouth and watch me pay. And then you can have some water and we'll go home. My waffle, your loss."

Of course, noticing how often real life was creeping into Adam's sleep talking, I couldn't help but try and experiment to see if I could willfully influence the content of STM's musings. Much to Adam's annoyance, I started addressing STM just before we went to sleep. The first time I tried this, I said, "Sleep Talkin' Man, are you in there? Listen, I'd really like to hear about manatees tonight, okay? Can you tell me something about manatees?" In response that night, I got:

> "'Sea cow this' and 'sea cow that.' Fucking bastard sea cows taking up all the conversation! It's MY turn to shine."

Sounds like STM resented my showing interest in any particular subject matter, rather than his mere existence. Touchy, touchy!

Another night, he further revealed his refusal to play by anyone else's rules. Just before we went to sleep, we were Skyping with my friend Jenny. Jenny, a vegetarian herself, signed off the conversation by saying, "Don't say anything mean about vegetarians tonight!" To this, we got:

> "You know, the world will be a much better place when we get to eat vegetarians. Furthermore, you get your five-a-day with one of those."

Apparently, STM doesn't take requests.

So while real life references do pop up in Adam's sleep talking—and make a nice addition to the egomaniacal self-praise, scathing insults, and inexplicable surreality, there's no mistaking it for reality. I can assure you, in his waking hours Adam knows how to share a waffle.

> "I wanted a shark with laser beams, and I got a manatee with a Maglite. For fuck's sake, get back in your hole and get it right."

ME: *You said, "I wanted a shark with laser beams, and what I got was a manatee with a Maglite."*

ADAM: The poor manatee.

ME: *Why? It's just a Maglite, he probably doesn't even feel it.*

ADAM: He can use it as a weapon as well. He can hit something with it.

ME: *A manatee would never hit anybody.*

ADAM: Out of self-defense, maybe?
They've got some weight behind them,
those manatees. They're slow,
they can creep up on you.

ME: *They creep up on you and just
snuggle you!*

ADAM: And then they kiss you with their
bristles. They can fight crime with
kindness and cuddles . . .

ME: *Oh, I like that.*

ADAM: . . . and then hit you over the head
with a Maglite.

"

If honesty's the best policy,
and the truth hurts, then you'd better call an
ambulance, 'cause you're not gonna like the
stuff I'm gonna fucking say.

♥

Finger painting's fun.
I need to get some more fingers, though.
Give me your fingers. Oh fuck it, I'll just
cut 'em off. You've got pretty fingers.
I can do some pretty painting with those
pretty fingers. Toes, I can use toes! Yay!
I'm sorted. Fingers and toes.

♥

Batman, why don't you sit down and
have a cup of coffee. I feel much
more comfortable calling in the
marines. Rubber-loving freak. And
as for you Robin, put some fucking
normal clothes on, dickhead.

"

"
The fruit flies have escaped!
Hide your plums! Satsumas flee! FLEE!
Oh, this is gonna be total fruitocide.
Fuck you, avocado, you're on your own.

,

You've got to save the curtains!
Save the curtains . . .
they hold so many secrets.

,

By the way, washing in rose water doesn't
stop you smelling like a piece of shit.

,

Goddammit! Where's my bazooka?
I put it on the ground, expecting it to
be there when I come back. Have you been
tidying up again, 'cause I really would like
my bazooka back. I don't know where you
put things. Jesus! Cannot just leave
anything alone, can you?
"

"You piss-flap motherfucking tosser!
No, I DON'T want to say anything else.
I think that sums you up enough."

•

Okay. Brace yourself. I can't be held
responsible for the fallout. You
ready? I Like You. Yeah, rock 'n' roll.

•

Shoot the fucking cellist.
Pompous cunt with his oversized violin.

•

I smell because I'm a PENGUIN
and I eat FISH, doofus!

It's Captain Fluffer!
Hero to teenage boys.

❦

Yay! It's my birthday today.
And you're going to give me presents.
Big fuck-off presents. Lots of them . . .
WHERE ARE MY PRESENTS?! Tossbag.

❦

I need a mask, nun-chucks, rubber suit,
shitloads of talc. Got list, will shop.
It's a superhero shopping list, shithead.

❦

God judges you. Well, I judge God.
This week: not bad! Still messing up in
the Middle East, though. Sort it out.

"

Being overweight's hard work.
You should applaud my determination.

♦

No. I don't talk about
the jellybean incident. That's one wound
that'll take too long to heal.

♦

So what, vegans are healthier
and live longer.
Let them be the last ones living
on a dead planet.

♦

Kiss my imperial ass,
you Jedi muthafucka!

"

**Yeah, I love you.
But in a not-really kind of way.**

❦

**Right then:
After dinner, it's butt shaking time.**

❦

Don't you dare get me up early.
Especially to get vegan fucking cupcakes.
What ever possessed you? Eat some mud
instead. Tastes the same, and I bet there's
more nutrients in it.

❦

I'm gonna make you laugh so hard your
holes will leak. Wet pants time!

"

Here's what we do:
We wrestle back control from
the pirate gerbils, and the seas
will be ours! That's right,
you nasty little sea fluffies . . .
Shit, they're stashing the treasure
in their cheeks! Come on!
Time for some plundering!

●

I've gotta have more people in my life
that love my elbows.

●

Step outside, needle-dick bug fucker.
There's a whole world out there,
and they just want to hate you, too.

"

My foot is itching to have a date with your ass. And if you really want, we can make it a double date. Your face and ass, my fist and foot.

From now on, you will OBEY ME. 'Kay?
Now just nod your head once for yes.
That's good.

Vaginas have brains. Oh, they've worked me out. And they don't like me. Oh!

My donkey. That's MY donkey.
Get off my donkey!
You know, you're not some
superstar donkey jockey. Piss off.

Is it a bird? Is it—Oh. It's Koala Man. Asleep in the branches again, and probably pissed. What an embarrassment to superhero-kind.

Letter to Sleep Talkin' Man

A while back, I was the guest of a dear old friend. One night I got out of bed, walked downstairs, strode right past my host sitting in his easy chair reading, and took off out the front door. His pet raccoon, Racket (evidently figuring I knew where the action was), followed me.

I walked right across his yard and breezed through his neighbor's front door, the raccoon hot on my trail, went straight to his refrigerator, snagged a pound of bacon, turned around, and headed back outside.

On the way out, I patted the head of his neighbor—fifty-nine years old and a Vietnam veteran—and said, "Good boy! Play nice with the ladies," then strolled back up the hill, gnawing strips of raw bacon, the raccoon plodding along in my wake.

"Worms for the weak!" I crowed as I headed back into my friend's house and climbed upstairs with my prize. I crawled back into bed, and Racket the raccoon and I shared a salt-pork feast.

To add to the delight of this scenario, I was bucknaked the entire time.

Of course, I only found all this out when I was told about it the next morning, everybody rolling on the floor laughing over it. As proof, Racket had the trots from eating too much bacon. I didn't remember a damned thing.

Stony

Saluda, NC

> "Oh, the penguin parade. Splendidly turned out, in his black-and-white coat and beautiful red-and-orange beak. And following him is— It's dressed exactly the same. No imagination, these puffins. Boring fuckers, really."

There are certain motifs—perhaps you've noticed?—that crop up again and again in Adam's sleep talking. Of course, there's the high incidence of insults, and the relentless ego-wanking you're now well familiar with. But beyond those, there are some preoccupations that suggest intriguing things about Adam's subconscious.

> "You tap-dancing hamsters. You rock my world, with your little blurry feet."

Perhaps the most prevalent topic among Adam's midnight mutterings is animals. Whether it's gangster manatees, hippo ballerinas, heat-packing badgers, or chickens in fetish costumes, animals get a lion's share of late night airtime around these parts.

In the universe of STM, an animal may evoke joy, as in the quote above. Often, however, the creatures that populate STM's world are clearly jerks:

> "Check out that crab,
> and his oh-so-sideways attitude.
> Dipshit. No one's impressed."

And occasionally, he comes out with something that would elicit a cringe in any animal lover . . .

> "Snail fiddling is not an occupation I'd be
> proud of. You dirty fucker."

. . . Let's just say I'm pleased not to be a fly on the wall for those dreams.

It's no surprise that beasts and critters feature so heavily in STM's world, as they're such a big part of Adam's waking life. The animal kingdom —whether creatures of air, land, or sea—is a passionate interest that Adam and I share. We watch more National Geographic Channel that can possibly be healthy; we can both identify every one of our sixteen goldfish by name; one of the criteria for our next home will be a yard that can accommodate chickens and goats. Even our wedding date was scheduled around the arrival of my dog Molly from the United States, so that she could walk down the aisle with Adam's children.

These days, most of our travel focuses around animals as well. As I mentioned earlier, our honeymoon was spent volunteering at an elephant sanctuary in Thailand. You might be thinking, hey, doesn't everyone want to celebrate their new life together by getting up at 6:45 a.m. and working their asses off shoveling elephant shit in 95-degree heat? But for us, it was perfect, absolutely magical. The following year, it was a monkey sanctuary in Ecuador, and then sloths in Costa Rica. We're hoping for manatees next. I can't wait to hear the sleep talking gems those inspire.

> "Stop the cows!
> They're licking everything!"

Cows in particular come up a fair amount in Adam's sleep talking. It's no wonder that these animals, peaceful and unassuming as they are, would have been branded forever onto Adam's subconscious; he spent some months of his eighteenth year tending, herding, and milking 250 of them. This was during the year Adam lived in Israel, just before university, the year we met. In between falling in love, Adam's kibbutz kept him busy working in the dairy.

On Adam's first day on the job, he got thrown right into the thick of it. After a cursory tour of the dairy, his new boss told him that he needed to deliver a baby calf from a cow in natal distress.

"Here's what you'll do," said his boss. "Stick your arm up the cow to the shoulder, and wait for further instructions. Hold on, I'll be right back." Adam was so eager to seize this once in a lifetime experience, and to show on his first day what a great worker he was, he figured he'd get started and be ready when the boss returned. He carefully rolled up his sleeve as high as it would go, and pinned the tail out of the way with his left hand. Then, making a little beak of his right hand, he gently inserted it into the cow's vagina. He eased his arm in up to the shoulder, determinedly thinking of warm rice pudding all the while, and waited for his boss to come back and express his surprise and gratitude for this new young worker with such initiative. The boss came around the corner, took in the scene, and sauntered over to Adam and the cow. Adam prepared himself for the approving pat on the back that was surely coming. The boss reached up and clapped his hand on Adam's shoulder. Shaking his head, he said, "If I knew how eager you were going to be, I would have told you to put on one of those." Adam followed his gesture down to find a pair of two-foot-long rubber gloves sitting at his feet.

It's not just animals that get the benefit of STM's consideration. We hear a lot from him about food as well. The roles that edibles play in STM's universe bear a striking resemblance to those of animals. There's the good:

> "Chicken soup. Get down with that funky stuff. Mmmmmm."

The bad:

> "Don't put the noodles and the dumplings together in the boat. They'll fight! The noodles are bullies. Poor dumplings."

Aaaaaaaaand the inappropriate:

> "Well, that certainly was an interesting use for a banana. Now go wash your hands. And burn the banana."

And, just as we hear a good bit about cows, there are, of course, favorite foods that put in frequent appearances. Like cake.

> "It's cake o'clock! All day long."

Adam talks in his sleep about cake an awful lot. It makes sense: Adam and I both LOVE cake, and we're willing to go the extra mile for the ultimate cake experience. Cake is awesome! Do you think it was easy finding someone in the UK to make red velvet cupcakes for our wedding? Red velvet may be the in thing now, but no one in England had ever heard of this American southern specialty a couple of years ago. We combed the countryside in search of someone who could make these moist, delicious, shockingly hued confections. But that's just the sort of dedication it takes to be a true cake enthusiast.

Adam's children reap the benefits of his love of all things sweet and spongy. Every year, for each of their birthdays, Adam makes an elaborate cake in the theme of their current favorite interests. This past year, his fishing-obsessed son had a guy being towed underwater by a fishing pole, having just been pulled off a jetty by a shark. His horseback-riding daughter had a horse show ring, complete with pony and tiny hand-made jumps. Last year, they got an aircraft carrier and a rainforest scene, respectively. If ever there was proof that Adam is nothing like his alter ego, I think this little habit is it.

> "It's taken years of cake abuse to get this body into the peak of physical fitness."

STM's passion for cake is equalled, luckily for him, by his utter lack of regard for the "thinner is better" mantra we're bombarded with all the time. Often, he appears to be triumphantly throwing an attempted insult right back into someone's face:

> "Me, fat? Think again, titty-fuck. I taught my muscles to be in a zen-like state of relaxation. Permanently."

Other times, he sounds downright proud of his extra poundage:

> "They're not love handles. No. I've got love impact protection barriers."

And yet, despite his refreshing comfort with his own apparent portliness, STM doesn't seem to have any qualms about using others' weight as a source of insult:

> "If you want me to be honest, then I have to say, your ass makes those jeans look small."

Of course, as STM insults go, that one was nothing. STM's passion for the put-down is unequalled and unignorable. One gets the sense that STM can muster contempt for just about anyone. And yet, I can't help but notice that there are certain groups for which he reserves especially vitriolic disdain.

> ## "I'll tell you what's wrong with this country: our proximity to France!"

The French are one such unfortunate faction. In fact, the poor French get it from both sides, waking and sleeping: awake Adam enjoys the occasional jibe at their expense as well, which is his birthright as an Englishman: just think of Monty Python and every other English comedy ever made. In fact, I've noticed that disdain for the French—ironic or genuine—seems to be the norm among people from all nations around the world, except the French (who, the stereotype would have us believe, feel disdain for everyone else, so I suppose there is a sort of symmetry there).

Surprisingly, evidence demonstrates that STM is open to having his prejudices challenged by Adam's waking experiences. When Adam and I spent two weeks volunteering at a monkey sanctuary in Ecuador, we shared a house with two fellow volunteers, both French, with whom we got on brilliantly. The walls of our little domicile were pretty thin, and I'd warned our housemates that they might hear any manner of weirdness coming from our room throughout the night. You can imagine my chagrin when, about a week in, STM loudly pronounced:

"Hmm, I like you. But you're French. So don't tell anyone I said it."

And yes, our French friends caught every bigoted syllable, as evidenced by the great guffaw I heard on the other side of the wall.

I'll tell ya, though, the French have it easy compared to the poor vegetarians.

"Vegetarians will be the first to go. That's my plan. Vegans haven't got a hope. 'I eat air, I'm so healthy . . .' Bollocks!"

No other individual or group faces quite the depth and breadth of abuse that STM bestows upon vegetarians. I have no idea where this comes from. As the cliché goes, some of our best friends are vegetarians! In fact, once upon a time, Adam himself gave vegetarianism a go. Doesn't every lefty teenager go through that phase? It lasted until the very next time his mother served roast lamb with mint jelly.

Finally, we couldn't possibly talk about common STM themes without considering the repeated appearances of all the typical heroes to little boys:

"You know, you can't be a pirate if you haven't got a beard. I said so. MY boat, MY rules."

Pirates, ninjas, superheroes . . . these are the figures that populate STMville. I guess since Adam's subconscious offers a playground where he can be whoever he wants, it's no surprise that he delves back into the fantasies that grown-ups are forced to leave behind, but probably still bubble away quietly beneath the conscious surface.

There are so many other motifs that crop up regularly in Adam's sleep talking—the combative nature of vegetables, the importance of good hygiene, the Jesus debate (son of God, or the first zombie?)—but I think I'll stop here and leave you to discover them for yourself.

THE BODY BEAUTIFUL

8 "I haven't put on weight. Your eyes are fat."

7 "My ass and my personality are the same thing. Huge and in your face."

6 "Right, like I joined to get fit. I joined for the mirrors!"

5 "I'm not fat. It's just my awesomeness swelling up inside me."

4 "I've got muscles. They're just sleeping. Don't wake them. Let them sleep."

3 "Yeah, I'm shapely. I'm a great big gorgeous shape and loving it."

2 "It's not sweat. It's my aura glistening."

1 "Well, so what if you call me fat. I'll forget you even exist the next time I see a doughnut."

Letter to Sleep Talkin' Man

One night, I awoke to the cries of my hungry one-year-old daughter. That was a nightly occurrence at the time and, as usual, I woke my husband Eric and asked him to go get her for me so that I could nurse her. It took a few tries to get him up, and by then, the baby was screaming loudly. Eric left the room and was gone for a really long time. All the while, the baby was still screaming, and I started to wonder what he could possibly be doing. After a good five minutes, he finally returned to the bedroom, but instead of the baby (STILL wailing in the nursery), he was carrying a winter coat. He handed it to me and said, "I remembered the coat, but I forgot the other thing."

Melanie M.
Washington, D.C.

"

**Ninjas in stilettos. Fashion assassains!
Not so stealthy but oh so stylish.**

"

> "You can't drop them. You can't set them on fire. You can't feed them to crocodiles. You can't let them play with fireworks. I mean ... kids: what the fuck?!"

I met Adam's children during my first visit to Adam in London, just three weeks into our rekindled relationship. Adam's son was seven, his daughter four, and from that very first day, they were wonderful with me. They are delightful, funny, easy-going kids, but also, I think that seeing their dad happy (which they hadn't seen much of lately) was all it took for them to immediately accept me as part of their family. Since then, it has been smooth sailing all along.

Adam's children get nearly as much pleasure out of STM as I do. We don't let them actually read or hear most of it for themselves, mind you. After all, they are only eight and eleven years old, and we do have some respect for age-appropriateness. But, of course, they know about their dad's funny little habit, so most mornings when they stay over they come bounding in, asking "What did Daddy say last night?" and I feed them a sanitized version, replacing "bum" for "ass" and "jerk" for "motherfucker." And let me tell you, it sometimes requires a fair bit of creativity to take STM's latest and greatest and translate it into something that's palatable to youthful sensibilities (and won't scar them for life). Other times, there's no sanitation in the world that will render something shareable with young ears: "How can I tell you you're as welcome as a twenty-eight-day-old used tampon infested by maggots without offending you?" just doesn't have a G-rated version.

Of course, some quotes are kid-suitable, and we happily let the children listen to those for themselves. The lengthy, wacky animal ones go down especially well. Here's the kind of thing they go crazy for:

"Hey, look at me! I just made bumble bee pajamas. They're so cuuuute, with their little leg holes. This one's bright fuchsia with some black spots. I think that's my favorite. And this one's got a night cap that fits right over the antennae . . . WING HOLES! I forgot to put wing holes! Oh, well. '. . . and all the honey was oozy woozy, sticky and gooey, but it tasted good.' Awww. Bed time story for tired little bumble bee. Go to sleep, bee."

It's a blast to watch their faces go from eyebrow-knitted puzzlement to wide-mouth amazement to unbridled glee. Kids haven't yet learned the regrettable skill of moderating their reactions, and it's such a pleasure to see it all hang out like that.

As delightful as it is to play this stuff for the kids, it still comes second to my marathon STM-sharing sessions with my brother. Running his own theatre company on top of a full-time job chairing the arts department at a private high school keeps Jason frantically busy, and he doesn't have much time to peruse the Net for sheer pleasure. When we Skype, we often get on the blog together and go through all the entries that he has missed. As we make our way through the quotes, he throws his head back and howls, his body contorts with laughter, he drops his forehead to the desk and pounds it with his palm. I've even seen the occasional tears of glee. It's embarrassingly gratifying to be the bearer of such merriment to my big brother.

It happens that Jason's young high school students have also discovered Sleep Talkin' Man. Many a time he has arrived at a drama club meeting to find the kids doing dramatic readings of STM's latest zingers. There he is, sitting with these fifteen-year-olds whose impressionable minds he is tasked with helping to shape, as they proclaim:

> ## "It's the soup!
> ## It tastes like rancid cock butter!"

He knows he's supposed to stop them, but all he wants is to cackle along. And it can't help that they all know that it's come to them care of his little sister. AWKwaaard!

Speaking of awkward, try sitting next to your mom as she listens to a recording of your husband saying:

> ## "From now on, papaya shall be known as cunt fruit. Nasty cunt fruit. Mushy and smelly cunt fruit. You don't like the word, don't make me say it again."

My mother—a woman more likely to exclaim "fiddlesticks!" than its four-letter counterpart—is surprisingly enthusiastic about Sleep Talkin' Man, particularly when you consider that it involves hearing all manner of obscenity from the man who married her daughter. What a sport.

"

Heaven for a depressed masochist
is an ice cream headache.

•

Ladies and gentlemen, please remember
to put your oxygen mask on first,
followed by your favorite child.

•

Oh, I could be rummaging around in here for
ages, I'm never going to find some zebra ears!

•

Ugh, I know you. You're alwayson the corner
of Fuck-off and Cunt-bag.

•

I'm so sorry about the Pop-Tarts.
It really should never have happened.

"

> I want to dance in the rain but without the getting wet bit.

> Just put the fucking cow's head on the pavement and walk away. Leave it alone, stop playing with it. It's just a head. Ooooh, it's got it's eyeballs in still.

> Hey, who put my elbows on backwards?! That's not fucking funny!

> Right. I've had enough. I'm splitting you two up. You over there and you are going all the way over there. I tell you, you've got to be really fucking quick and hard on these chinchillas. Take no prisoners.

"

Don't judge me.
Anybody can fall in love with semolina.

Stop throwing mangoes.
You're going to take somebody's eye out,
or worse!

Put Mr. Squidly down!
How DARE you try and milk him! Come on,
Mr. Squidly. Let me put you back in your
tank. Aw, it's okay. Why don't you hug my
arm. Yeah, use all your little tentacles.
There there. Everything's gonna be okay.
He's only a douche.

Listen, it's not as if I put ear wax on
my penis and shouted 'snake warts!'
OK?

"

"

Yeah, OK, you're sorry. You're French,
you've got to be fucking sorry!
Mange tout twat.

●

I can't believe you went to pick up a turkey
without introducing yourself first.
How rude of you. How presumptuous.
A turkey has its own mind. Be kind.

●

Seriously, I can open my mind
and empty it of everything and
still do menial tasks. Picture that.
I am the perfect husband.

●

Sure you're beautiful.
But when you crap you smell like every
other asshole.

"

189

" Leave my gnomes alone. They're MY gnomes, living in MY house, doing MY gardening, and they're happy. Look at their fucking smiley faces. Can't you see how frickin' happy they are? Who are you to judge me?! Go on, gnome, cut the grass. Good gnome. Good gnome.

❦

Stupid fucking fizzy fish. Never liked them. Have some of that, you sugar-coated cunts.

❦

If you're looking for sympathy, go get a fucking dictionary. You'll find it between 'shithead' and 'syphilis'.

❦

It's your hair. I'd like to see it on your head, not on the side of the fucking bath like a dead mouse. "

**You find me attractive?
Well, congratulations. You've now
joined the rest of society.**

•

No, don't laugh at my goose.
Come on, goose. Oh, this is going to cost
me a fortune in therapists.

•

The ravioli's plotting something.
Always hiding his agenda. Stick with fusilli.
Really trustworthy.

•

Okay everybody. It's time for some whale
song. Get ready: mmMMMMMMMMmmmm,
MMMMmmmmm, mmmMMMMMMmmm,
MMmmmMMMMMmm . . . .Oh, I'm filled with
so much humpback happiness right now.

> "Don't judge me by the friends I keep. No, no, no. Judge me by the enemies I have slain!"

As you can imagine, Adam's sleep talking can turn some situations rather awkward. We now warn in advance anyone who will be sleeping within earshot. This is a lesson that we learned the night Adam yelled out, "SOAPY FUCK-ING TIT WANK!" loud enough to be clearly heard by the nice older couple staying in the bamboo hut right next to ours on our honeymoon. We even mention it not only to kids who are coming for sleepovers with Adam's children, but also their parents. We'd rather not have their kid come home telling their parents that they heard in the night, clear as a bell, "If Santa doesn't bring me my Xbox, he's a dead fucking fat cunt!" Even if that is a sentiment a ten-year-old boy can get behind.

Once I created the blog, it was no longer just friends and family who had access to the deep, dark inner musings of Adam's subconscious, but potentially anyone with an Internet connection, and there have certainly been consequences to that. For example, the agency that Adam was working for was not at all comfortable when their account manager attained his fifteen minutes of Internet fame. In no uncertain terms, he was instructed never to reveal the name of the company in interviews about the blog, nor to reveal to anyone in the industry that he was Sleep Talkin' Man. He was the Clark Kent of sleeperheroes! "What if a potential client sees the blog, doesn't like it, and takes their business elsewhere because of it?" they reasoned. Personally, I think they missed a trick—Adam is in a creative industry (film advertising) and though I'm no expert, it seems just as likely that Sleep Talkin' Man could have attracted clients, rather than repelled them. Nevertheless, Adam kept these two areas of his life distinctly separate as was requested.

A few weeks later, purely by coincidence, Adam was headhunted to interview with another agency. He arrived, and soon found himself sitting across a conference room table from the director. "First things first," Tony said. "Are you Sleep Talkin' Man?" Given the directive from his current agency to maintain the secrecy of his alter ego, Adam wasn't sure how to respond. But, he figured, might as well get the truth out there early. Adam's cautious assent was met with great delight, followed by an announcement to the entire office. As it turned out, he had quite a fan club among the staff. Of course, they hired him for his qualifications, but their love of STM certainly didn't hurt! And his alter ego has proved, after all, to be a great ice-breaker with clients.

STM: MANAGER EXTRAORDINAIRE

10 "Hey, don't say anything. Why don't you put it in an e-mail, then I can ignore it at my pleasure."

9 "Sure you've got a job here. If you wanna work somewhere where you're NOT FUCKING WELCOME."

8 "Your job is to be ignored. Nobody's to acknowledge you whatsoever. You should be good at that."

7 "Wow. If you really think that was a good idea, maybe your mum should rethink having YOU was a good idea."

6 "You call that work? I call that a fucking fatal accident. I'm a witness to your carnage!"

5 "Can everybody come into the boardroom please! Not you, you don't work here any more. Bye bye."

4 "I'm tired of looking for the solution to this problem. Look for someone to blame instead."

3 "Your blue sky thinking is blighted with dark clouds of piss-poor ideas."

2 "Don't come in to work tomorrow. In fact, don't come back at all. Basically, I don't want you around, cause you're—I'll keep this simple—a cock. A small, pathetic, flaccid, looking-at-your-shoes-constantly kind of a cock. Okay, bye-bye!"

1 "When it comes to being told what to do, I tell, you do. Got that, dickhead?"

When I was in my late teens, my mother woke one night to a rhythmic banging. She quickly identified it as the sound of my headboard hitting my wall repeatedly. Figuring I was about to wake the rest of the house with my latest romantic conquest, she hot-footed it to my room to stop me in my tracks.

What she saw when she opened my door was not what she expected. There I was, sitting on our new rowing machine, ON MY BED, covered in blankets, rowing for all I was worth. I only wish I had used it half as much when I was awake!

Les P.
Newcastle Upon Tyne, England

"

Oh, tremors! Quick, under the table! Leave the goldfish though. Little shit. Let him suffer. Thinks he's all high and mighty 'cause he's got a castle.

❦

Everybody wriggle. Everybody wriggle. It's maggot mayhem!

❦

I will NOT wear my lobster suit and dance in the street. Not even for rhubarb and custard. Go away and leave me alone. My bee costume is waiting. Bzzzzzzzz.

❦

When I'm king of the coalition, nobody's going to be able to poo at work, ever. A dirty waste of smelly time.

"

"

**I only have eyes for you . . .
and here they are,
in this lovely presentation box.
All for you.**

❦

**Five balloons. Got to be five.
No point going to the disco without five.**

❦

There's this guitar riff stuck in my head.
Doo doo doo doo-doo, doo doo doo doo-doo.
Whoever wrote it . . . is a cunt,
because it's stuck in my head. Bastard.
I'll stick something in his head.

❦

I'm gonna mess you up so badly, Stick Man,
that when I'm finished with you, you're just
gonna be a scribble. Yeah!

"

"

Oh, get up and wash your shadow.
It's filthy. Filthy!

❦

How much for the frog?
No, that one . . . No, THAT one.
Goddammit, how much for the frog,
that one with the tail? . . .
Well excuse me!
How much for the lizard then?

❦

The joke's on you, God.
I'm free will in action.

❦

Mazel tov, cuntbag.

"

"

That's right.
Crème anglaise, motherfucker.

●

**Yeah, you can keep looking at
my ass as I walk away.
It's having a spectacular day today.**

●

All things considered, you are more
attractive than a monkey's vagina
. . . on heat.

●

Happy Birthday! It's a dead puppy! . . .
Now listen, you: You didn't specify a live
puppy, you just said you wanted a fucking
puppy! Jesus you're spoiled.
Now go take it for a drag.

"

"

Stop telling everyone we're friends. Don't amplify my shame.

●

Oh, calm down.
You don't wanna get him started.
You DON'T want a kick-boxing hamster
on your case.

●

Right. So, I've punched the singing
telegram. Now all I need to do is throw up
on the cake. Party time!

●

Opposable thumbs! That was nature giving
man the opportunity for twenty-four-hour
fondling. Thank you!

"

"

You're full of horseshit!
Like bullshit, but it's dressed up
to be prettier.

●

Since when did my underwear look good
on you? Take it off. Take it off your face.

●

Don't leave the duck there.
It's totally irresponsible. Put it on the swing,
it'll have much more fun.

●

You are as much use as a cup full
of monkey spunk at a monk's retreat.
And just as pointless.

"

"

I'm scared by the power of your vagina.
It can control whole armies.
No, nations. The world!

❦

Fuck. Tits! Shit, I brought the wrong shit!
It's shower-hat Wednesday, not cracker-
pants Thursday. Bollocks! I really like
shower-hat Wednesday. I've got the best.
It's so pretty ... these cracker-pants give
me nasty chaffing. ... Oh, I shoulda put
margarine on them! Lubricate the cracker
bits. Yes, yes, yes ...

❦

Llama clouds! Mmmmm.
Great big bug-eyed fluffy ones ...
Oh great! Now they spit. Bollocks.

"

> "Ah, glass. My nemesis.
> One day I shall beat you."

If ever there was a time that STM spoke out for Adam, it was the very moment that he uttered the quote above.

Adam has a rather unfortunate relationship with glass. Or you might say that his head has a rather passionate relationship with glass, as the two don't seem to be able to stay away from each other. In our apartment, the windows and glass doors are all ornamented with oval-shaped grease marks from Adam's forehead. Ditto the driver's seat car window. I don't know what the problem is, whether it's a matter of faulty depth perception or a complete absence of kinaesthetic awareness, or what. There just seems to be a magnetic attraction between Adam's face and any vertical sheet of glass.

Adam's daughter, sympathetic to her father's handicap, has taken to making all sorts of pretty little art projects that she can tape to our garden door at Adam's eye level. You know, like people put up for birds, so they don't crash into windows? What a sweet kid. His son, on the other hand, delights in trying to lure Adam into these minor mishaps. "Look, Dad, a pygmy hippo in the garden!" "Wha—" *CLONK.*

Adam's tendency toward self-injury is not limited to interactions with glass. He is outlandishly accident prone; every corner, table leg, door frame, and household item from can openers to light bulbs poses a potential hazard to his person. Being Adam is a truly perilous pastime.

Take the other night. We were out to dinner with the editor of this very book. Adam left the table to go to the bathroom, which, like many London restaurants, was in the basement. When Adam got back to the table, he was rubbing his forehead. "Oh no," I said, "What did you do?" It turned out that, while he was going down the stairs, he'd been distracted by a sign and hit his head on the low ceiling. What did the sign say?

"Watch Your Head."

Adam crashed his moped on an open road with no one else around. He went on a school ski trip, fell off the chairlift on the first ride up, and spent the rest of the trip laid up in the hotel. The first time he ever threw a boomerang, it worked.

Is it any wonder, then, that I've forbidden Adam to participate in any of the following activities:

- Skateboarding
- Snowboarding
- Paintballing
- Riding motorcycles
- Riding horses
- Contact sports
- Any activity that ends in "-gliding" or involves a racquet (Except badminton. Badminton is allowed. I'm not a monster.)

Adam's propensity for damage is so powerful that it can infect those in his orbit as well. For example, when Adam was nine, his parents brought home a new kitten from the SPCA. Little Suki was having a delightful little kittenhood, until she decided that her favorite pastime was to climb up Adam's body and perch on his shoulder like a parrot. One day, Suki took a tumble from her special spot. You would not expect this to be a problem: cats always land on their feet, right? But, don't forget, this is ADAM's cat, falling from ADAM's shoulder. Poor Suki landed on her side, breaking one little kitty leg.

Because Suki was still a kitten, the vet made her a tiny cast with a little extra space to allow her leg room to grow. For the next six weeks, all anyone heard around the house was pat pat pat THUNK... pat pat pat THUNK. Her relationship with Adam was never quite the same.

Fortunately, unlike STM, who consistently presents himself as God's perfect gift to the world, Adam has always had a knack for being able to laugh at himself. This ability has allowed him to translate a lifetime of mishaps into an endless supply of amusing anecdotes. It's certainly been a skill he's gotten lots of use out of.

Letter to Sleep Talkin' Man

Late one night, I awoke to find my computer-geek husband palpating my abdomen as though he was carrying on some sort of medical exam. This is the conversation that ensued:

Him: *(with authority) Mmmm, yes, I see, uh-huh, this is not good.*

Me: *What's going on?*

Him: *(continuing his physical exam) Well, it's really clear, Sweetie. This thing has to come out.*

Me: *(a bit alarmed) I don't think so.*

Him: *Well, there's no doubt, it's gotta go or you won't make it up this hill. You need to make it to the top.*

Me: *Well, I think I'd like a second opinion.*

Him: *No need for a second opinion. I can do this, it's easy, just let me find my Leatherman.*

In case you don't know, a Leatherman is a utility knife that some men wear on their belts.

Lucky for me, he fell back to sleep.

Sarah B.
Homer, AK

"

My badger's gonna unleash hell
on your ass. Badgertastic!

"

> "At the rate I'm going, I'm gonna need another dozen turkey bags. Oi, back in the bag. Back in the bag, damn you!"

ADAM: What's a turkey bag?
What, is it an American thing?

ME: *Some people do cook their Thanksgiving turkey in a special kind of bag in the oven, 'cause it keeps the juices in.*

ADAM: No, we don't have that here.

ME: *This sounds like you're talking about live turkeys.*

ADAM: Maybe I'm putting children into turkey bags.

ME: *It seems a good idea, quieting them down that way when you're tired of them.*

ADAM: Yeah. Instead of the naughty chair, get in the turkey bag! This is Parenting 101.

Frozen yogurt is ice cream for pussies.

*

Ground Control, this is Weasel. I'm all
ready to go. Just waiting now . . . itchy fur,
itchy fur . . . Ground control, this is Weasel.
I think I've got fleas . . . begin countdown.
. . . Beep. Beep. Beep. Beep. Beep . . .
Ground Control to Weasel,
I hope you're house trained.

*

**You're a happy little soul, humming away.
If you like humming so fucking much I'll
staple your bitching lips together. Shut up!**

*

**Smell my hands. They smell of bacon.
Go on, smell. It's okay, it's kosher.
Mmmm, bacon hands.**

**You're right, elephants in thongs
are not something you see every day.
Enjoy it.**

❦

**You cannot go to college unless
I teach you shit. So you can take that
'those who can't do teach' crap,
and shove it up your flabby ass.**

❦

**You're pretty.
Pretty pretty pretty pretty pretty. . . .
Now fuck off and be pretty somewhere else.
I'm bored.**

❦

**I'm a fucking goose, dickhead!
Duck, indeed.**

"

Jellyfish are attacking.
Everybody grab your ice cream guns.
Let's get those little things!

♥

Reach for the stars!
You're gonna have to, no one on this planet
wants anything to do with you.

♥

Bad germs, bad germs.
Whatcha gonna do?
Whatcha gonna do when I
throw bleach on you . . .

♥

You've got a nice face.
But your personality is one huge cock block.

"

"

The cake! It must be psychic.
It's using Jedi mind tricks to make itself
irresistible to me. This is the cake I want.

Run along, run along. Next!
I love being a judge on
Nobody's Got Talent But Me.

Thank you for dinner.
Now if I ever say that something tastes
like the anal evacuation of the recently
deceased, I'll have a frame of reference.

Here I am: Captain Yeast Infection!
Making you uncomfortable no matter
what you wear.

"

"

I bet you wish you could wear
a onesie and look this good.

,

Next time I lend out my chest hair,
it won't be to a dick face like you.
You've got it covered in lots of stuff.
It'll take ages to clean up.
One hair at a time.

,

It'll be less painful to put my tongue
through a cheese grater and lap up vinegar
with the remnants that are left in my mouth,
than to accept a dinner date with you.
I think you can take that as a 'no.' Bye-bye.

,

So, you want a super massivo
decaf low-fat stupido motherfuck
cappu-fucking-ccino?
OK. Coming right up, cunt.

"

"

I swear, if I ever have to listen to you,
I'll claw my ears off, and fill the bloody
gaping holes with decaying afterbirth.

Do I love you a little or a lot? Hmmm?
No, it's a lot. It's your shithead inner self
that only loves you a little.

Back off Robin.
Batman is my bitch now.
You're just a bitch's bitch, bitch.

You know what I'm gonna do?
I'm gonna make turd biscuits.
It's like a shit sandwich, but for kids.

"

"

It's not so much a wave of indifference
that washes over me when I look at you.
It's a fucking tsunami. Fuck off.

•

Nature pooing gives a whole new dimension
to life. Mmmm. Now that's freedom!

•

Stop hitting me! You're always hitting me!
Hmm, maybe I shouldn't be a boxer.
Postman, I'll be a postman, yes.
No one hits a postman . . . except if you're
a boxer. Yeah, maybe I won't be a postman.
I'll just be a god. Much easier.

•

I love you. No, sorry, I missed
some words out. I love it when
the world shits on you. Yeah.
That's the full sentence.

"

"

Jesus needs me.
If I don't believe, he don't exist.

•

I'd like to apologize
on behalf of my mother.
She didn't mean to call you an asshole
. . . she meant to call you a cunt.

•

I want to see the piglets.
Let me see the piglets.
Why can't I see the piglets?
Ohhhh, piglets!
. . . Fuck they stink!
I want to go home now.
Stinky fucking piglets.

•

Yeah, you lay down,
I'll get the tennis racket.

"

> "What the fuck are you doing here?
> Piss off back to the sick part
> of my imagination."

As we've already established, Sleep Talkin' Man
is such a separate entity to Adam's waking self
that Adam himself even refers to STM in the
third person. And I'm not just talking about
those glaringly obvious digressions in personality: Adam's good-natured humility to STM's
egomania, Adam's generosity and kindness to
STM's insulting irreverence. There is so much
more evidence than these basic traits alone to
suggest that a wholly separate being is unleashed
when the lights go down.

For example, people always want to know
whether Awake Adam is as hilarious as Sleep
Talkin' Man. Well, Adam is a pretty witty guy,
and even has the odd moment of comic genius.
But does he have that pitch-perfect punch that
STM manages with such frequency? Not even
close. Adam couldn't come up with this stuff

awake, certainly not this much of it, if he tried. And, believe me, he does try. But where STM rarely delivers a clunker, Adam is never afraid to go for the eye-roll-inspiring, low-hanging fruit joke. You know, the kind of puns that only dads make. Here's a typical example, from one of our early morning conversations:

KAREN: *Last night, you said, 'Unless you've got a chicken shoved up your vagina, you can shut the fuck up. Alright, class, pop quiz!'*

ADAM: We don't say 'pop quiz' in the U.K.

KAREN: *Really?*

ADAM: Yeah. If anyone said 'pop quiz,' it would be a music test . . . you know, pop . . . it would be a music—

KAREN: *Groaaaaaaan . . .*

STM: 1, Awake Adam: 0.

But that's far from the end of it. There are all sorts of words and phrases that STM uses that would never come out of Adam's mouth. "Douchebag" is a distinctly American insult, never used by Adam in his waking life or, for that matter, by any self-respecting Brit. And, I'm sorry, but "hot dang" in a mildly posh English accent sounds ridiculous. Or how about this:

> "Seriously man. I was mega wacked-out. True, blood. Peace out, muthafuckaaah."

Huh?! Is STM moonlighting as a blaxploitation film actor?

Adam's nighttime natterings are peppered with all sorts of un-Adam-like, un-English stuff like this. And I know exactly what you're thinking: no, he didn't pick it up from me. Of course, like most Brits, Adam has seen his share of American movies. I guess STM's definition of cool is Quentin Tarantino.

More amazing, though, is the fact that STM sometimes comes out with words and phrases with which Adam, to the best of his conscious knowledge, has no familiarity. One night early in his sleep talking history, Adam muttered,

> "Dance for me, go on. Oh, you were! I thought you were having a spaz attack . . . Doofus."

When I told his what he'd said, he chuckled, and then said, "Umm, what's a 'doofus'?" And how about this:

> "We gotta get out of here. Oh! There's a giant aye-aye coming, and he's pissed!"

Neither of us had any clue what he was talking about. After trying out a variety of spellings in Google, I eventually determined that an aye-aye is a type of lemur. Even after this discovery, Adam had no recollection of ever having heard of such an animal.

But it's not just the odd word or phrase that distinguishes STM from Adam. STM has even

228

been known to mutter in foreign languages that Adam does not speak. Here he is dabbling in French:

> "Touché, mon petit frère. Now it's MY turn to fuck you up."

I actually posted this on the blog as "Touché (blah blah blah in french)..." because neither Adam nor I, listening to the recording, had any clue what he had said. Luckily, a few readers e-mailed us with the proper transcription and translation, "my little brother."

Now, how about a bit of German:

> "Ja. Bitte schoen, mein lieblich. Mit knackwurst. Mmm-hmm. Und strudel. JISM!"

At least in this case, I knew enough of the language to make sense of it myself. And of course that last word, bellowed out with great enthusiasm, is pure English/American slang (and, incidentally, not an ingredient typically used in German cooking).

All that said, I certainly don't believe that STM has access to knowledge that is completely alien to Adam. Sleep Talkin' Man can only be a figment of Adam's subconscious, so all this stuff must be in there somewhere. Some psychologists believe that a function of the subconscious is to sort relevant information from irrelevant, so that the conscious mind knows what to focus on. Perhaps this is all random mental detritus that just got caught in Adam's subliminal filter, only to emerge at night. It's either that, or STM is, after all, a totally separate entity, a night spirit who roams the earth—fleeing rare lemurs, dueling French swordsmen, swearing like a longshoreman— periodically possessing Adam's sleeping body only to share his experiences with the living. I'd go with theory number one.

My cousin often sleepwalks. Late one night in a hotel, he walked out of his room on the sixth floor with his ice bucket, took the elevator down to the lobby, walked to the ice machine, and filled up the bucket. Then, he got back in the elevator, went up to the FIFTH floor, and pounded on a stranger's door thinking it was his room. When they opened the door, he handed them the bucket of ice, walked over to the refrigerator, opened the fridge door, peed into the fridge, crawled into the stranger's bed and went right back to sleep.

I can just imagine the review of the hotel that person wrote!

Kristin B.
Alberta, Canada

**Hey, don't creep up
on me like that!
Superheroes are wound
super-tight.**

"

**I'm losing faith in humanity,
one faked orgasm at a time.**

♥

You never take my balloons out for a walk.
They need some fresh fucking air,
take 'em outside this time. And on a long
walk. They like the sun. Don't take them to
the park. I don't want them on the swings,
they're too little for swings. And the
roundabout will make them sick, just take
them on a nice walk. See the duckies.
Bye balloons!

♥

You know, it's a human race. And you lost.

"

"

Oh shit. It's the munchkin sing-song.
Kick 'em in their mouths,
the tuneless fucking freaks.

●

Butt cheeks ahoy!
There she blows!

●

Mine's a potato martini.
Serve it cold.

●

Well, if I'm the douchebag,
you're the contents, Titfuck!

"

"

I got my big meeting today.
The one where I stand up and say,
'fuck this shit, I quit!'

⁋

God, you whine like whale song.
But a lot less eerie and beautiful
and more, well, fucking annoying.

⁋

Elvis is dead. He is dead, I tell you!
Buried and oh, just a little bit smelly.
Bad burger muncher.

⁋

Whoever invented calories is gonna get
their face fucked with ice cream cake.

"

"

Of course it's your tomato.
You can do what the fuck you want with it.
Except have sex with it.

♥

Fucking unicorns.
'Oh, I've got a horn!'
So fucking obnoxious.
You're just a horse with
a party hat, dickhead.

♥

My butt cheeks are for squeezing.
Go on, take a handful. Take two.

♥

Oh, great. Now you're older than Jesus.
Your one great achievement in your sad
fuck of a life.

"

"

I'm sorry, but not knowing what
a horcrux is is a deal breaker to me.
Deal with it, muggle-fucker.

●

It's a turd. I wrapped it, and put a ribbon on
it, and I'm giving it to you. See? I give a shit.

●

I hate bubbles. Oh sure, they're nice
and pretty to begin with, but then the
rainbow colors just go POP! Instant
downer. Motherfucking bubbles.

●

I wanna be a bumble bee. But not a slutty
bumble bee, going around, pollinating every
flower it sees. I'm a one flower kind of bee.
A monogamous bee. Monogabee! That's me.

"

Cream cheese to the moon, motherfucker!

♥

Oh, I've got the tears of a hippo ballerina on my arm. Don't touch. Don't touch! Sparkly. Oh so delicate and sparkly! Oh, sparkly.

♥

Ladies and gentlemen, in the event of sudden change of pressure in the cabin, you may wish to scream into your masks. SCREAM BITCHES.

♥

Red panda . . . blue panda . . . blue panda . . . Green! Panda, stop changing colors. Someone go and get Chameleon and find out what the hell's happened here.

"

Of course the mermen are all pissed off. The mermaids are all bitches. And they can't sing . . . . How do they have sex?

♥

Mary had a little lamb. I ate it. Mary's sad. Stupid whiny vegetarian!

♥

Don't stop me. I need to put this on my Santa list before I forget. It's my Santa list! You know, an Xbox 360 for me . . . and a cock slap for you. I'm so excited.

♥

Seriously, there's nothing like a good ass-licking before you go to sleep.

"

"

Ooooh!
I'm gonna play chubby bunny until I puke.
Yeeaaahhhh.

♥

I'll buy the cow and put it on the roof.
High-rise farming is gonna be
MY invention.

♥

Yeah, my balloons! Watch them dancing
in the sun. Aren't they—Fuck!
Fuck, you little fuck! Your kid's a tossbag
for popping them. Tossbag kid.

♥

Ohhh, Snuffleupagus.
You're such a hairy cunt.

"

> ## "Today's a bad day to be my underpants, that's for sure."

ME: *. . .You also said, "Today is a bad day to be my underpants, that's for sure."*

ADAM: Oh dear . . . it's never a bad day to be my underpants. In fact, underpants are queuing up just to be worn by me. Did you know that? I open my drawer and they're all screaming, "Me! Me! Wear me!"

ME: *Awww.*

ADAM: And when I close the drawer again, they all go, "Oh nooooo! Not the daaaark!"

ME: *Ohhhhh.*

ADAM: But it's okay, 'cause the socks keep them company.

ME: *How do you choose?*

ADAM: Well, it's like a lucky dip, I just close my eyes and rummage around . . . It's like the claw.

ME: *Well, no, because usually the claw doesn't get anything.*

ADAM: I'm the good claw. But then, there's one pair of socks, an old pair, wizened and frayed at the back of the drawer. He's been there for a long time, he's see it, been there, done it, and he keeps telling stories to all the other socks to keep them going. Some day the old pair of socks will come out and never come back.

ME: *No, leave him there!*

ADAM: He's going to the sock drawer in the sky.
                   That's their goal in life. It's a place
                   where the drawer never closes, and
                   they never have holes, and they're never
                   frayed, and they're never jumbled up on
                   top of each other, they're folded nicely.
                   It's the sock drawer in the sky.

ME: *Mmm, sounds really nice.*

ADAM: And they're always in a pair, they're
                   never single socks.

ME: *If the washing machine eats a sock, but
                   you keep the other sock around for a
                   while thinking maybe you'll find the
                   sock, but then you send that sock to the
                   sock drawer in the sky, they're reunited?*

ADAM: Yes. Definitely. Unless they're bad socks, and they go to sock hell. And they're permanently stuck in the washing machine.

ME: *What does a sock have to do to go to sock hell?*

ADAM: It's one of those socks that constantly twists itself around on your foot when you're wearing it, so it gets really uncomfortable and the heel gets twisted round on the side of your foot. Or it keeps falling down, and comes off in your shoe. Those are bad socks.

ME: *Yeah.*

ADAM: And they go to sock hell, and they go on a spin cycle for eternity. But every sock strives to be a good sock and go to the big sock drawer in the sky.

> "You certainly are incredible. A perfect example of genetics gone wrong."

I've heard it said that sleep talking may be genetic. Perhaps there is something to this. Adam's mother doesn't quite talk in her sleep, but she does have her own special sleep behaviors. She sometimes has nightmares, in which she begins to scream. Now, in her dream, it's a horror-movie scream. But the sound that actually comes out of her sleeping body can only be described as someone trying to do an impersonation of a siren, a rapid "WOO WOO WOO WOO WOO WOO!" Not a nice awakening for anyone else sleeping in the house. She's also been known to carry out brief, amusing actions from a dream. One night, for example, Adam's father woke up to find her sitting up, punching at the air. She woke up suddenly and when he asked what she was doing, she replied "I decked the au pair."

I've recently started to wonder whether sleep talking is contagious. In these past couple of years, since Sleep Talkin' Man emerged and became a hot topic of conversation among family and friends, we've had a number of people in our lives—Adam's father included—start talking in their sleep for the first time! Sounds to me like their subconsciouses are jealous of Adam's subconscious.

"
Vampire penguins?
Zombie guinea pigs?
We're done for . . . done for.
"

> "Imagine waking up next to you every day ... One chunder-bucket moment after another."

If you were previously unfamiliar with the term, you have probably now pieced together the meaning of "chunder-bucket" for yourself. If not, think barf bag and you'll be on the right track. This sort of utterance makes me so thankful that I'm married to Adam, who courted me with the utmost determination to make me his wife, rather than STM, who associates married life with perpetual vomiting.

From the moment that he came through the airport doors on our first reunion, Adam was intent on marrying me. He was not perturbed by the cynics of the world, including the one that was, it would seem, lying dormant in his own subconscious.

Adam started dropping marriage into the conversation from the second day of that first visit (you remember, the one where we were just getting together for a friendly cup of coffee). He didn't talk about it, in a serious heart-to-heart. Nor did he toss it out there in jokey, offhand comments, as though testing the waters. He simply referred to our future life together as a part of normal conversation, as a foregone conclusion, with utter confidence in the rightness of it.

But I needed a bit of convincing. On that first visit, I wasn't sure how I felt: I had a lot of baggage from our shared past that needed unpacking and discarding. I also needed some time to reconcile Adam at thirty-four with the eighteen-year-old boy with whom I had first fallen in love. It was a lot to work through. So whenever Adam confidently asserted the inevitability of our future nuptials, I changed the subject without pretense of subtlety. I wasn't denying, but I certainly wasn't confirming.

Adam returned home from our six-day coffee date without any proclamations of love or assurances of a future from me. We had three more weeks of nightly Skyping, and then I, still quite unsure of my feelings, flew over to his stomping grounds to bring in the New Year. It was on December 30, on the escalator of the Tate Modern, that the last of my doubts fell away and I knew I was unquestionably in love with Adam. The certainty hit me all at once, the same way I instantly felt his presence on that first night in Israel half of our lifetimes ago.

It was two weeks later, during his second visit to the States, that Adam next mentioned our future marriage. This time I responded shyly with something like, "That . . . could . . . maybe . . . happen." Adam is a guy who can spot an opportunity. In that moment, he got down on one knee, took my hand, and asked me to marry him. Well, first he had to brush some popcorn and candy wrappers out of the way, seeing as we were sitting in a Brooklyn movie theatre waiting for *Juno* to start. It may not have been every girl's romantic dream, but for me it was perfect.

I didn't feel like I could truly get engaged without my parents ever having met the man in question, so we decided that we were engaged to be engaged and thus, to fully win my hand, Adam fearlessly, and awkwardly, faced each of my family members, one at a time. We scheduled a trip down to New Jersey.

First stop: Dad's house. My father lives in my South Jersey hometown, on a little body of water that, when I was a kid, was referred to as "the sandwash" and was where teenagers went to have sex and do drugs. Now, "Shadow Lake" is the idyllic setting for a handful of upper middle-class homes of doctors and lawyers.

It was a crisp Saturday afternoon, the day after we arrived. At this point, we had been there less than twenty-four hours, eight of which had been spent sleeping. So Adam had racked up, let's say, ten hours getting to know my Dad. The lake was frozen, and Adam and I were down on the jetty skipping rocks across the ice when my dad wandered down with some stale bagels to feed the ducks (given the passion that the Shadow Lake birds habitually show for bagels, I'm convinced they're Jewish ducks). Since we were just a few hours shy of leaving, it seemed like the right time for me to make myself scarce so that the men could talk.

As Adam tells it, they had been tearing off pieces

of bagel and tossing them onto the ice for a while when he took a deep breath and said, "Skip, there is something I want to ask you. I would like to—"

"Yes!" my father jumped in with alarming eagerness, "You can marry my daughter!"

Adam, who had prepared himself for a serious moment, was flustered. "Oh, uh, well, OK then," he said, and took his bear hug like a man.

Having now covered the principal topic at hand, neither of them had any idea what was supposed to happen next. So, they went back to tossing bits of bagel to the ducks. The problem with this course of action was that the ducks had never, in fact, come to the jetty at all. So there they were, a man and his future son-in-law, loitering awkwardly in the middle of an ever-increasing semicircle of baked goods, desperately wishing that the woman they both loved would come back and rescue them.

I, having never been in this situation, was feeling extremely shy, and had taken to hiding in the house. Adam and I had never discussed what was supposed to happen after he asked, and I had no intention of going back out and risk interrupting the manly heart-to-heart that I assumed must define such occasions.

To their credit, they stuck it out until not a crumb of bagel was left in the bag. They gazed a few moments longer across the expanse of bagel lumps until Dad said, "Hmmm. I guess she's not coming back." And they trudged back up to the house.

Next, we drove up to my mom's in central Jersey. Again, we gave it about two meals' worth of getting-to-know-you time before Adam brought up our future plans. Luckily, two meals gets you surprisingly far in getting to know my mom. She's immediately familiar and welcoming, the kind of parent that all of her teenage kids' friends called "Mom." Even so, it's nerve wracking to put in a request for marriage on first meeting with any parent. Sunday afternoon rolled around, and Adam knew he couldn't put it off any longer. My mother was rushing around the kitchen, cooking for a dinner party. It wasn't ideal, but Adam feared it would be his last opportunity to catch her alone. I was waiting just around the corner from the kitchen, out of immediate sight, but within earshot. Adam, in his awkwardness, expressed his intentions in ridiculously posh, outdated terms. Something like, "Patti, I'm sure that you've become aware of my intentions toward your daughter, and I would like your permission to have her hand in marriage." It must have sounded fantastically British to my mother, like something out of a Jane Austen novel. She paused in her chopping, the knife hovering above the carrots. "Are you planning to treat her wonderfully and make her happy?" she asked. "Umm . . . yes? I am," Adam awkwardly affirmed. "Oh," she said, "well then, she's all yours." They shared a nice welcome-to-the-family hug—although, she still had the knife in her hand, so I guess it could have gone either way. As Mom returned to her chopping and Adam came around the corner from the kitchen, I saw him do a Rocky Balboa over-the-head double fist pump of triumph.

Two down; one to go.

In hindsight I can appreciate that I made an error with my brother. With my mom and dad, I waited until they met Adam before there was any whisper of marriage. Adam is totally guileless and (in my totally unbiased opinion) utterly loveable, and anyone who saw us together instinctively knew that we belonged together. But in my brother's case, I just called and told him that I was engaged. His lukewarm, skeptical reaction was not all that I would have hoped for.

Put yourself in Jason's place. Your sister tells you that she is going to marry a foreigner who only six weeks ago she saw for the first time since having her heart broken by him a decade and a half before. Add to this that you generally believe this sister to be impulsive and not always possessing perfect judgment, on top of which you're an emotionally cautious kind of guy to begin with. You can imagine, then, that Jason was a little suspicious. I believe that, in short, my brother figured this was a guy gunning for a green card. "I'm sorry I can't respond with the hoots of congratulations that you were probably hoping for," he said. "That's OK," I replied, "you should respond however you feel." I was confident, you see, that he would thaw the moment he met Adam.

So, parents covered, I took Adam up to Boston. On our second night, we were out at a pool hall when my brother tricked me into giving him some man-to-man time with Adam. "Tamar wants to talk to you about something," he said, handing me his cell phone with his girlfriend on the other end. I took it across the room, where I could hear better. A theatre director, Jason always knows how to inject just the right amount of drama to communicate his point effectively to his audience: He bent down, aimed carefully, took his shot, righted himself, planted the end of his cue firmly on the ground, and pinned Adam with an accusatory gaze. "So," he said evenly, "What is it that you want from my sister?"

The content of what followed is known only to Adam and Jason, but Adam must have given a convincing answer, because by the end of the weekend, they were delighted with each other. A year and a half later, it was Jason who officiated our wedding with that same sense of dramatic, but this time it was suffused with joy and love.

Of course, at this point, none of us had met STM.

Letter to Sleep Talkin' Man

I'm not just a sleep talker; I'm a sleep doer. Many a morning I've woken up to find my roommates snickering into their coffee, tears running down their faces, all too willing to regale me with stories of the crazy things I said or did the night before. I'm apparently a fount of information in my sleep. For example, I knocked on a roommate's door the other night and when she answered it (knowing full well who'd be on the other side) I informed her that "only the male crickets creak." I'm sure that trivia will come in handy some day.

Once, I tried selling Girl Scout cookies door to door down the hallway and apparently got frustrated when no one appeared to be home at the bathroom door. I yelled, "Fine, fartknocker, I'll just cheek you sideways then!" and stomped back to bed.

My sleepwalking habits can especially be a problem when we have unsuspecting overnight visitors. One time, my roommate's parents were visiting. Apparently, I marched into the living room where they were soundly asleep on the air mattress, jumped on the couch, pointed to the door, and proclaimed (in a crackling witchy voice), "The Gate, The Gate! Don't go beyond the Gate! That way lies madness (then in a deep voice) AND CERTAIN DEATH." Then I "doom doom doooooomed" and stomped back to bed. I swear I don't remember a thing!

Shawna S.
Sunland, CA

**Hey logic, you can suck my balls.
I wanna do some shouting.**

♥

**Look at the size of your bath.
I can pee in it and you'd never notice.**

♥

Why must I choose? Dog or fish.
Dog or fish. . . . Fish . . . ARGH,
I get it wrong every time!

♥

**Damn it, I'm gonna be late. I've run out of
nipple glue! Always at the worst times.**

♥

Methinks it's time to go naked native.
It's a shower cap and singlet for me.

"

Oh, put the phone down. . . . No, you put it down first . . . No, you. . . . Just put the ass-rimming pig-fucking mother-shit fuck phone down! Jeeesuuuusssss!

•

Gaffer tape.
Oh, it's such sexy sticky stuff.
Rip and stick, rip and stick. Ooooh.

•

Look at me. That's what I call rapture.
Who needs the fucking end of the world?
Judgment day, my ass.

•

You have the genetic disposition of a dipshit. It's quite simple to trace back. One of your ancestors must have risen to the highest rank of codpiece.
Stupid fucking wanker.

"

"

The little people are taking over!
Better break out that secret
stash of rainbow pencils.
That'll keep them busy for a while.

♦

The zero is the same. It always will
be. It never changes. Zero is zero. . . .
One? Huh, he's just a lonely cunt.

♦

If I can't have legs,
none of you can have legs. Simple as that.

♦

I can juggle babies. It's the baby juggle!
It's just so tricky when they're on fire.
One up, two . . . oops! I need another baby.
Another baby!

"

"

GYM: nothing more than Goddamn Yucky Masochism.

♥

I don't wanna go down the slide. It's too high. Far too high. Wheeeeeeeeeeee! . . . I feel sick now.

♥

It's not about believing or not believing in God. No, no, no. It's about not giving a crap.

♥

Fluffy bunny + twitchy nose + big ears = great stew.

"

"

I'm gonna kick you so hard, your nuts are gonna look like Christmas balls. Now give me back my wings! And the tiara.

❧

I'll have you, Blackbeard. And then I'll have your beard. Mmmm, I love stealing beards. Be the Pirate Beard of the Bearded Sea. Arrrgggh. And all of you follicly challenged people can be on MY crew. And everyone with beards will quake with fear!

❧

OK: Jump position! Goddamn it, why do the jellyfish always get it wrong? "Oh, I've got my own built-in parachute." So fucking what, get the basics right, dickhead. Otherwise, you're out of the free-form team. Bastards.

"

"

Of course I'm wearing my thunder pants. It's Thursday. Friday is frilly pants day. Don't forget!

•

I'm not angry with you. I'm just pissed off you were classed as a viable embryo.

•

Tell me, what are vampires wearing this season? Anyone? Anyone?!

•

I know I said I love you, but I love me more. Accept your position in my life.

•

Stupid-fucking-cunty-bollocks-expialidocious

"

"

Oh, hamsters don't give love like guinea pigs. No, no. Guinea pigs are a love package, all wrapped up in little squeaks.

I've got a new show, and you're gonna be the star. It's called, "People I Like to Throw Under the Bus."

I call this cake Death By Icing.

Ninjas, they're just pussies in pajamas. Samurais, now that's where it's at. *They're the fucking bollocks.*

This is a totally sparkly tiara moment!

"

> "Now it's time to suspend reality . . .
> from its fucking neck."

"Just imagine," I often think to myself excitedly, "what Adam would be capable of if he could summon STM at will. WOW!" Other times, I think, "Just imagine what Adam would be capable of if he could summon STM at will. UGH!" In a sense, STM demonstrates the extreme best and worst that Adam has to offer. He's Adam at his most clever and creative, and he's Adam at his most narcissistic and unscrupulous. Both of these extremes, I think, can be explained by the same phenomenon: lack of inhibition.

Because, really, that's all STM is: what remains, or is allowed to emerge, when all of the inhibitions that govern so many of our behaviors are stripped away. First of all, you get purely unfettered creativity. With no fear of failure, no concern for consequences, no judgmental internal editor, the mind is free to fly and allow for the best of itself.

But let's not forget, inhibitions do us at least as much good as harm. Sympathy, empathy, and morality—in short, our conscience—act as inhibitors against an entire focus on the self. STM is all about me me me. He takes what he wants, does what he wants, says what he wants, with no limitations imposed by the consideration of others. Such freedom allows for a capacity for cruelty that would make most of us shudder.

And this seems to bring us full circle. Who else do you think of when I say "unfettered creativity, self-absorption, capacity for cruelty"? Yes, children! Here we are, back at the theory that STM is basically a child, with an adult's vocabulary and knowledge of the world.

I can't help but wonder what sort of world we would live in if everyone's inhibitions were stripped away. Imagine our entire global society made up of STMs. Would we be so much more scientifically, technologically, artistically, culturally advanced than we are now? Or, would we each be so busy wallowing in our own awesomeness and joyfully delivering cutting put-downs that we wouldn't have gotten around to inventing the wheel?

TOP TEN
STM Insults

10 "It's easy to confuse your ass and your face.
They're basically a hole that spews shit."

9 "It's amazing how you can smell so bad,
but still be alive."

8 "Why don't you stand in fuck up corner.
You can stay there 'til, I don't know,
I-don't-give-a-shit-about-you 'o clock."

7 "I'd rather be afflicted with permanently infected
hairy ass grapes than be subjected to another
minute of your drivel."

6 "Listen to you? I'd rather listen to the
sound of me sucking out the juices of a
corpse through its anal sphincter."

5 "Scientists in the future will completely
struggle to work out how you were ever
classified as an intelligent life form."

4 "You know, I really think you should audition for Britain's Got Talent . . . at being a cunt. You're a shoo-in for a winner."

3 "Watching you think is like watching a cat shovel shit with two broken paws. Painful, but I just can't stop watching."

2 "You, sir, are the used sock on a teenager's floor. Nothing more."

1 "I'm sure you'll be thrilled and ecstatic at the thought of going out with me. But look at it from my point of view: Leeches attached to my testicles and a mass of flesh-eating caterpillars and ants over the rest of my body, gorging themselves on my flesh. A slow and ultimately painful death. So I hope you can see my point of view. It'll be a no to going out with you. Now run along and poison somebody else's life."

"

Harder is NOT a good safe word.

"

> "That's the problem with the food cupboard. There's too many places for zombies to hide."

ME: *You said, "The problem with the food cupboard is there's too many places for zombies to hide."*

ADAM: No offense to STM, but zombies aren't clever at hide and seek.

ME: *Hmmm, that's true.*

ADAM: They like to surprise, but they can't because they've got their arms stretched out. It ruins all the hiding places. They go behind a door, but their arms are stretched out, so the door can't close. They go behind a curtain, but you see the lump of the arms. Plus the fact they're always groaning about brains, so they can't be quiet. They could be under the bed, and you just walk in the room and all you can hear is, "braaaaaaaiiiiins." It's like, "Found you!"

Letter to Sleep Talkin' Man

It's around two a.m. and suddenly I am wide awake (or so I think). I sit up quickly, squint my eyes to see where I am, and then start walking around the room picking things up and examining them. Then I turn to my partner and, holding up an object for him to see, say, "Look at this, we have exactly the same one at home!" He tells me to get back in bed, that I'm sleepwalking, but I ignore him. "No, REALLY, look at this, they have all the same shit that we do . . . and they put it in exactly the same places!"

Matt C.
Warwick, RI

> "I got to make myself some gin. Lots and lots of gin. Gin helps . . . everything."

One might think that alcohol would bring out the STM in Adam. Both states—asleep and inebriated—are times when people tend to be free of inhibitions, so it would be logical to assume that the same subconscious personality might emerge in Adam during each. Not so. Drunk Adam is cheerful, soppy, and downright adorable.

As I've mentioned, a side perk of having a recorder going every night is that I occasionally catch stuff other than sleep talking, such as Adam on an inebriated roll. Here is an excerpt from a typical Drunk Adam monologue, in which I was already in bed and he's just come back late from an evening out with new clients. You'll need to imagine this in an English accent with wild swings of modulation and volume:

"I have been told by somebody I have met FOUR OR FIVE HOURS AGO what a wonderful amazing woman I married." (I ask, "Well, how would they know?") "Because I TALKED about you FOR HOURS! I TOLD them how you make me FEEL, what you DO for me, how amazing you are with OTHER PEOPLE, and ANIMALS, and you're just YAAAAAY! . . . And, ummm . . . what was I talking about? . . . "

And so on, for the next twenty minutes. This is a pretty good representation of almost every conversation we have on the rare occasions when Adam is drunk.

And yet, there may very well be a germ of commonality between STM and Drunk Adam. Case in point:

Gin, so extolled in the quote above, played a noteworthy role in our wedding. It was the night before the big event. The last of our out-of-town guests had drifted off to their hotel rooms, I'd sequestered myself with my matron of honor to paint my nails and do other pre-wedding girlie things, and as the midnight hour struck, Adam had the sudden nausea-inducing realization that he had not given a moment's thought to his speech for the big day. He dashed to his room and grabbed some hotel stationery and pens, and went in search of a bright comfortable place to work. Ah, the hotel bar, perfect! Before he knew it, the friendly bartender had placed before the groom-to-be a nice fresh gin and tonic, on the house. This turned out to be the first of many free-of-charge G & T's the bartender would bestow unto Adam that night. Getting himself steadily sloshed, Adam delved, he delved deep, for the perfect words with which he would convey to all of our loved ones the overwhelming joy and passion in his heart on this most meaningful of days. And thus he began writing his wedding speech in the only writing implements he had been able to dig up, his five-year-old daughter's coloring markers. Bolstered by gin—and the ability to change hue every time he needed fresh inspiration—Adam composed the most touching, most hilarious, most goddamn kick-ass speech I have heard at a wedding, ever. Segueing seamlessly between heartfelt displays of emotion and parody of his loved ones (his opening, for example, is a reference to my well known, over developed passion for organization; my wedding planning spreadsheet was a fourteen-tabbed display of shock and awe), he had everyone gasping with laughter, and crying into their napkins. It went, in part:

Having lived with Karen, it is impossible not to pick up on her many intense and infectious traits. For me in this moment, it would be, most notably, the spreadsheet. At first viewed with dim regard as the tool of necessity for the gray accountant, I have since been known to stare in awe at the filtering of data, the formation of formulas, and the ebb and flow of our ever-mutating to-do list. And so it came to pass that my speech was born in a spreadsheet.

The speech, as my father explained to me, should be simple. Mention your wife, thank your mother. Mention your new in-laws, thank your mother. Mention your kids, thank your mother. The other important thing to remember is to avoid the minefields of inappropriate comments. So, in this spirit, I will try to avoid bringing up that mother-in-law is an anagram of Woman Hitler, and how excited I am at having new in-laws, especially ones that live 3,500 miles away.

Patti and Skip: Today your daughter has taken on a husband. Before you run out of the room screaming with unbridled joy at your release, I want to thank you. Not simply for today, but for your continued help, support, and advice. I only knew you for, quite literally, five minutes when I asked your permission to marry your daughter. Little did either of you know, this would lead to her immigration to a tiny island of ex-colonial powers and strange accents. For that, we must thank the Department of Homeland Security.

Mum and Dad: How can I ever thank you enough. Ah yes, by moving out of your home. Seriously, you have been an incredible support to me. When faced with an adult child moving back with a fiancée in tow, numerous children sleeping over on a regular basis, losing the sole use of mum's car, and feeding the five thousand, never once was I questioned, never a raised eyebrow, always an "of course" and "what more can I do." I salute you.

Karen: It amazes me how hard this part of the speech is, to put down in words how I feel when I look into your radiant eyes. To describe how my heart races when I see you, how my arms feel unfulfilled unless they hold you, is almost impossible. Your laughter fills and thrills me. I can go on, and many of you will admit that I do go on when I talk about Karen. Seventeen years in the making, and finally we are together. For more on how I feel about Karen, please visit www.mywedding.com.

It is at this time I am reminded that no man is truly married unless he understands every word his wife is not saying, and that love is accepting that your loved one's sound advice is just 99 percent sound, and 1 percent advice. My darling, I know you are right . . . MOST of the time.

(By the end of the book you'll understand that reference to the role that U.S. Homeland Security played in my immigration to the United Kingdom.)

I can't, however, say that Adam is ALWAYS at his sharpest when he's been at the bottle. A few months after our wedding, we went off on our delayed honeymoon—volunteering at the aforementioned elephant sanctuary in Thailand—which provided me my first opportunity to personally meet Drunk Adam in all his glory. Having spent the better part of the day constructing a barbed wire fence to house the sanctuary's crazy pony, he figured he was deserving of a quick drink in the local bar with his fellow builder. Thus was Adam introduced to Thai whiskey.

Two hours and some unknown number of shots later, Adam stumbled back to the sanctuary and up the steps to our hut just before dinner, giggling his head off. With a bit of cajoling, I convinced my grimy husband that he had to take a shower before dinner. I got him undressed, and sent him off for his wash. What followed was a trip into the absurd straight out of STM's most random moments.

The sanctuary bathrooms were a row of ten-by-ten tiled rooms, each with a hip-level tap, an enormous plastic bucket (the sort you might put your bags of garbage in for collection), and a large scoop with a long handle. The idea is to fill the bucket from the tap, then dunk the scoop in and repeatedly douse yourself with water.

I was lounging in our hut when I heard a cry of, "I'm stuck in the bucket!" Over and over, "I'm stuck in the bucket!" I ran over to Adam's shower room to help—only to find that he had, naturally, locked the door. Damn. I dashed into the adjacent shower room and clambered up onto the top of the toilet so that I could peer over the wall. It seems that Adam had a notion of taking a bath, so had stepped into the bucket and tried to sit down. Halfway there, he got himself wedged down. If there's one thing you can say for Drunk Adam, his eloquence in expressing his innermost feelings is surpassed only by his accuracy in describing his physical state. He was, indeed, stuck in the bucket.

There wasn't much I could do from where I was except laugh at his plight, and take a few choice photos. Once I was done with that, I climbed down from the toilet to formulate a strategy to free my husband from his plastic prison. But Adam was doing his own thinking. Soon, I heard, "thunk . . . thunk . . . thunk . . . CRASH . . . WHOOOOSSSHHhhhhh." Hopping back up onto my commode perch, I found Adam sprawled on the floor in a pool of water, giggling maniacally. Through his drunken haze, he had surmised that rocking back and forth until he toppled the bucket offered his best chance for freedom. It was certainly effective. And hilarious. Needless to say, I took more pictures.

Early the next morning, dragging his hungover self in for his work shift, Adam was greeted by the staff with gleeful cries of "I'm stuck in the bucket!" It turns out that sound carries pretty well across an elephant sanctuary.

I wonder what happens on a pirate's birthday. Does he get to choose who walks the plank? Hip-hip ARRGGHHH! Hip-hip ARRGGHH! Pirates are funny.

❦

Eyes down for a full house. I love drag queen muppet bingo!

❦

Hello conscience. This is my wife. She makes the decisions around here. Oh yeah. Don't get any fancy ideas.

❦

Garlic cheese! Double death to you, you lactose intolerant vamp man!

“

I can't drive with the roof down,
but where are we gonna put the dolphins?
They love to feel the wind in their blowhole.
Pffffffffffff, blowhooooole! Click click
click squeak, click squeak squeeeeeak.
Awwww, they said 'I love you.' I love you too,
dolphins. Just sit down in the back.

❦

The only sharp object I carry is my wit.
And I'm gonna cut you up, bitches.

❦

You know I love it when I hear those three
little words. Come on. You know you want
to say them to me .... Yeah. I AM amazing.

❦

You've got to respect people's beliefs.
I believe you're a miserable wank stain.
Piss off.

”

"

My computer needs more power.
Feed it chips. Lots of chips. With ketchup.
Not mayonnaise.

❦

Oi! God! Shut the fuck up and listen to me.

❦

I'm the pilot. It's my turn to fly the plane.
Give me a peaked cap and a stewardess.
Yeah, a stewardess. Whooossshhhhhhhh.

❦

**Assholes of the world unite!
And fuck off together.**

❦

You're about as welcome as anal leakage.
Now fuck off and infect somebody else's life.

"

"

I know it's a shame that when
I walk out of a room it gets just a little bit
darker and gray. It's a burden I carry.

●

You speak your mind, I punch your face.
I think it's a fair exchange.
We'll both be hurting.

●

I can't wear these pants anymore.
They're just too tight.
They're giving me cock cramp ...
FREEDOOOOOOOOOOOM!
Yeah, wiggle it. Mmm, feel that
swaying? That swaying is freedom!

●

You don't make any sense. You must be
part of the alien menace. Stop with the
retarded hand gestures already. Stop!

"

"

**I'm in the mood for kicking faces
and punching crotches.
Woo hoo!**

♥

That's the difference between you
and me. Your heart is filled with hate,
and mine is filled with kittens.
Mmmm, kittens. Meow.

♥

I would gargle contents of the assholes
of the recently dead than go out with you.
It's not a hard choice really.

♥

Yeah. It's a long journey to find
your soulmate. So here's a one-way ticket
to somewhere far away, now FUCK OFF!

"

"

I don't listen to the crap you say.
Why should I give a fuck about the shit
you tweet?

•

Graphic novels: They're just comics that
grew up, flipped you the bird, and waved
a hairy nut-sac in your face.

•

"Unique" and "special" are the two words
I'd use to describe you. That is,
if I'm not allowed to use "cunt" and "bag."

•

You're undiluted brilliance,
awesome to a point of purity.
You're— oh, sorry, I was projecting.
You're an ass.

"

"

Those ladybugs are racist cunts. Don't you dare invite them to the garden party.

🌢

I've just bought fifteen bags of this shit, and now you tell me you don't like gummies anymore? Bloody hell! What am I gonna do with this gelatin? I was gonna make the biggest fucking gummy bear in gummy bear history. It was gonna be gigummygantic!

🌢

Ooh! My balls are itchy. Have you got the cheese grater?

🌢

Where do you think YOU'RE going, hmmm? I knew it. The cupboard. You and your cupboard.

"

"

Your family, or a zombie horde.
Choose carefully.

"

> ## "Here's my CV. Why don't you just file it under 'Awesome'?"

I've shared my theory that STM's raging ego is Adam's way of compensating for low self-esteem in his earlier life. I've also divulged his accident-prone nature, which has left him with a lot of old humiliations for STM to churn through. If ever there was an experience in Adam's life that explains why STM determinedly and unapologetically punctures the veil of politeness, demands his needs be acknowledged, and lets his freak flag fly, it has to be this:

When Adam was in his early twenties, he managed to land himself a job in PR, at one of the biggest firms in England with ultra-posh digs in the famous Russell Square.

Adam was nervous and excited as his first day of work dawned. Decked out in his brand new suit, he boarded the crowded Tube and headed toward Central London. He arrived at his destination and, jostled in the rush hour crush of bodies, Adam slipped on the platform as he stepped off the train. With a gymnastic feat of contortion, he recovered without falling, sacrificing his back in the process; with a disconcerting wrench, his muscles lodged their spastic complaint.

"I'm not going to let a little spasm beat me," thought Adam. "This is the first day of a fantastic new career!" With that optimistic sentiment he shuffled out of the station and around the corner to his new office.

By the time he got there, Adam was in too much pain to appreciate the beautiful oak doors, or the tastefully luxurious reception area. Gripping the grand banister of the massive winding staircase, soaked with sweat from his agony, Adam dragged himself up two floors.

He lurched into the lounge and hung against the door frame to catch his breath. This was not the grand entrance he had imagined for himself. Spying the solid-looking wooden table across the room laden with a lovely coffee/tea spread, Adam staggered across, planted both hands flat on the surface, and remained as such—bent double, dripping sweat, seeing stars—desperately hoping for the pain to subside so he could start the day with a bang, rather than a pathetic whimper.

By this time, Adam's new colleagues began to turn up. Spotting an unfamiliar figure, a number of them came over, and, with all the hail-fellow-well-met one would expect at a swanky London PR firm, slapped him on the back with various versions of, "Oh, the new boy! Welcome!" After the third slap, Adam keeled over backwards with a loud groan, hit the floor, and there he remained, prone, like a tortoise that had been rolled onto its shell.

The crowd was getting thicker around Adam's prostrate figure as more and more of his new colleagues arrived. You would think that people would gather around to offer aid to the poor young man paralyzed with pain on the floor. However, this was neither of the two sorts of responses that Adam observed: there were those who came over to him, squatted down, and introduced themselves heartily, taking care to avoid any acknowledgment that there was anything amiss about the way in which this meeting was taking place; and there were those who simply stepped right over Adam to get their morning cup of coffee without giving him a moment's notice. The latter group included women in skirts, who preferred to walk directly over Adam's face rather than acknowledge the situation. Putting aside the great views he must have had—which I doubt he was in a mindset to appreciate—I can't help but feel so sorry for my poor future husband, too young and insecure to demand the attention appropriate to his mortifying circumstances. I only wish that he could have found a little spark of STM in himself then, to command the treatment that he deserved.

Instead, Adam continued to hope against hope that he would at some point find himself pain-free enough to bound to his feet, laugh the whole thing off, and begin his new career. That somehow this would all simply become one of those things that everyone jokes about years later during a round of golf or over drinks on the yacht.

Meanwhile, over half an hour had passed since Adam hit the floor. By now, word had spread around the entire company that the new boy has started, and if you want to say hello, you'll find him lying on the floor of the lounge. The ever-increasing agony in his back was rapidly expanding his definition of pain and, on top of everything, he now seriously needed to pee. It was at about this time that some kind, proactive soul finally hit upon the novel idea of calling for an ambulance.

After what seemed an age, two paramedics suddenly appeared in Adam's field of vision, and with them they had brought a seven-foot board to carry Adam out. They successfully swivelled the board under him with minimal movement of his back, strapped him down, and carefully made their way out of the lounge. When they reached the stairs, they appeared surprised to discover that they were on the third floor, and the only way down was a long winding staircase. "We're gonna need another crew," Adam heard, and he was hauled back into the lounge and unceremoniously dumped against the wall. The next thing he knew, the paramedics were being offered a lovely continental breakfast, while Adam had become part of the furniture.

Finally, the second ambulance crew arrived. At last, rescue! They burst into the lounge and rushed over to the breakfast table, where they began discussing with the first set of paramedics the finer points of their respective days thus far, occasionally looking over at Adam with an absent-minded smile and a slight wave of a croissant. After a bladder-busting amount of time, they categorically declared that there was no way he would be getting down by the usual route. After some deliberation, one of Adam's new colleagues thought to mention the service elevator. So Adam, strapped to a board, neck in a brace, was paraded through the entire office.

As you have perhaps gathered by now, these particular paramedics may have not been the sharpest tools in the shed. It took them five tries to realize that they would not be able to load Adam on his seven-foot board horizontally through the door into the five-by-five elevator. One of them suggested that perhaps they should stand him up. Ah, a solution! Except that they fed him in feet-first. Thankfully, he did not fit that way either, which saved him from riding down upside-down. It was decided that what was needed was a fire crew to hoist Adam out the window and down two floors, to the ambulance waiting below.

Remember, these offices were in Russell Square, a prestigious and bustling area of London. In order to get a fire engine into the square, it was necessary for the police to completely shut it down to all other traffic.

So, there's Adam, prone, strapped to a backboard with neck brace, his suited body drenched in sweat, his back searing in white hot pain, his bladder fit to burst at any moment, and now the cause of a complete shutdown of Russell Square. His embarrassment was at its breaking point, and as Adam lay there begging the floor to open up and swallow him, into his frame of vision loomed the faces of the four paramedics, now joined by two police officers, and—yes ladies— SEVEN firemen. It could have been the perfect cast for a bachelorette party. But instead, this was Adam's first day at his new job and, now, one of the worst days in his life. And it wasn't over yet. He still had to endure the ignominy of the journey down the fireman's ladder from the third-floor window into the closed-off square below.

Thus began, and ended, Adam's illustrious career in public relations.

DON'T MESS
WITH THE STM

10 "I'm gonna fucking tear you limb from limb, and use your arm like a loofa and your face to clean my crack and balls. Now just go away."

9 "That's the green one taken care of. Bring me the blue and I'll kick seven shades of shit out of it."

8 "You give me stress, anxiety, days filled with woe. I give you, I don't know, a kick in the fucking balls. I think that kind of makes it fair. Asshole."

7 "You take one of those knitting needles and put 'em in my neck once more, I'm gonna see to it that every time you blink, you're gonna be looking at your own rectum. Got it?"

6 "That's it. I'm going to have to call an intervention on your stupidness. I think it will take the form of a brick."

5 "Pee in my bed once, shame on you. Pee in my bed twice, I'm gonna rip out your bladder and use it as a football, you geriatric incontinent cock slap."

4 "I think it's time you stepped into my office. The office of my fist."

3 "You try feeding me any processed soya, you're going to find it very hard to wipe your ass without any fucking arms."

2 "If you don't shut your cake hole, I'm gonna put you into a food coma."

1 "This is a friendly rock. Let me rub it on your face lightly. Yeah. Now it's got your scent, it'll like you. Let me show you: Stand there, and I'm gonna throw the rock at you. Watch how it wants to connect with you, time and again."

Believe it or not, Adam was not my first experience with wacky sleep behaviors. My brother Jason is and always has been a sleepwalker and talker! One night, when he was about eleven, my mother heard a commotion coming from downstairs in the middle of the night. When she crept down the stairs, she discovered my brother kicking and yelling at the vacuum cleaner, which he'd dragged out of the closet. He nearly broke his toe! But think of the bravery, risking that precious digit to protect our family.

On another occasion, my mother heard yelling coming from my brother's room. She ran across the hall and threw the door open, only to find him using his left hand to bend his right hand back with such force that he was near to breaking his wrist. It turns out that he had rolled over on his right arm, and that hand had fallen asleep. Meanwhile, his left hand was feeling around, and came across a foreign hand (the right, numb one). Believing the dead hand to be a monster, he attacked!

About a year ago, Jason started having a recurring sleepwalking episode in which he believes there is someone outside the door of his house. He gets out of bed and creeps through his apartment, with the intention of sneaking up and throwing open the front door to surprise the uninvited guest. He's gotten further and further each time, most recently finding himself with his hand on the knob of the front door. He's very concerned that the most likely next step is his actually flinging the door open and leaping onto his front porch clad only in his boxer shorts.

There was one fateful night that Jason's roommate Jacob stumbled home at two a.m. from a night out drinking. He let himself in the front door, and made his weaving way down the hall toward his own room. Unfortunately, he was passing by my brother's room at the very moment that Jason, mid-sleepwalk, yanked his door open. Jason, for the first time actually finding the suspected stranger looming at the door, screamed at the top of his lungs, causing poor drunk Jacob to pitch in with with his own scream of terror. Jason slammed the door in Jacob's face, and screamed again. With this scream, Jason actually woke himself up, and gradually made sense of what had happened. Doubled over with his hands on his knees, heart pounding, he opened his door again to find Jacob, near to hyperventilating, propping himself up on the other side. Both gasped and wheezed for a while, trying to calm down, until Jason said, "Don't . . . ever . . . do . . . that . . . again."

"

I just don't like those German shepherds
and their achtung sheep.

•

I'm the epitome of seeing is believing.
Once you see me,
you'll believe there is a god.

•

Duh. They're deaf.
They can't hear me.
YOU'RE ALL CUNTS!

•

Life is precious.
I'm not going to just sit here listening
to your pathetic fucking dribble.

"

"

I've got a badger, a dog, a cat, and a sack. Now that I've got 'em you can fuck off. All mine.

♥

Hey! This is MY playground. These are MY swings. That's MY climbing pyramid. And that's MY springy elephant! THAT'S MY SPRINGY ELEPHANT! You crusty knob-end. Bog off! Leave this playground to the king of playtime! . . . Mmmm, they're all mine . . . . I need a push. I can't swing without a push. PUSH ME! Where is everyone? . . . Bastards. This is MY playground.

♥

I've never seen a baby pigeon. It doesn't make sense.

"

**Buffalo wings? Are you insane?
Those cows can't fly. It's a lie,
I tell you. A fucking lie.**

♥

I think you should sit down.
Surely your ankles can't take the weight.

♥

I understand, but things have changed now.
Ever since the Chocolate Bonanza.

♥

Shhhhh! Why can't you midgets talk more
fucking quietly?! I hate small talk.

♥

**Just the thought of kissing you
makes me want to take a vegetable
peeler to my lips.**

"

I'm bored.
Let's go and trip some old people.

●

I'm sorry, I tried. But liking you is just too
far outside my comfort zone.

●

Why aren't you making me warm, hmm?
Hmm? Why aren't you making me warm?
That's your one fucking job, to make me
warm, why the fuck are you not doing your
job? . . . Being dead is no fucking excuse,
you make me fucking warm!

●

Between die-you-cancer-upon-my-
life and I-couldn't-really-give-a-
crap-about-you lies your life story.

"

"

**Beer is from Mars.
Chocolate's from Venus.**

♥

I'm giving out tickets.
Five minute slots to stand next to me.
One at a time. Enjoy yourself.

♥

Why don't you stop looking for answers
when your questions have as much weight
as a turd floating out to sea.

♥

I'm talkin' about motherfucking cookies
and apple juice.

"

"

I need you to take this stapler and ram it into your forehead. No, it won't solve the problem, but it will make me happier. It's funny! Now go bleed somewhere else.

♥

Sure you can sit next to me. But you're going to have to be prepared to be eaten if we crash.

♥

There's a reason you're such an asshole. You just don't have to keep telling everybody about it. People will work it out for themselves pretty quickly.

♥

Oh, it's time I got a tail. Yeah, a real strong one. No, not for climbing, so I can wrap it around your neck and squeeze the living shit out of you. Maybe then I'll go climbing

"

"

No pens. There are no pens here.
I can't do any work anymore.
I'm in crayon heaven.

𐫱

It's science.
It's meant to confuse stupid people.

𐫱

Scales. Must have scales. And razor claws.
I want some feathers. And a goggly thing
on its head. Yeahhh. Dinochicken.
Awesome! I feel like a god. All right,
what's next? Guineapigasaurus. Bring it on!

𐫱

I wanna put a dog in charge.
They don't start wars. They just want love.
And to sniff bums. Yeah, sniff bums.
Sniff.

"

"
Squid wrestling:
all tentacles and no substance.

"

Over the past couple of years, we've received a number of fretful e-mails and comments from readers suggesting that perhaps Adam has a serious psychological disturbance, and that his sleep talking is the foreshadowing of an inevitable future display of shock and awe. These communications invariably conclude with a plea for us to get him to a doctor. For those that share this view, I hope I can put your minds at ease: we've been put in front of a number of sleep specialists (an unintended consequence of the media attention the blog received), and we've learned that what Adam does is not so uncommon, nor is Adam a particularly extreme example of sleep behavior. Though he does seem to be especially prolific and clever, there are loads of others out there.

But for those of you who are still sceptical of Adam's mental stability, or those who, like me, are simply curious about the science here, I've invited psychiatrist and sleep specialist Hugh Selsick to tell us what this sleep talking thing is all about. Hugh, take it away!

A Word from our Friendly Neighborhood
Sleep Specialist

Sleep talking, or to give it its scientific name, somniloquy, is something of an enigma in the already enigmatic world of sleep disorders. Given the huge response that Sleep Talkin' Man generated when it went viral on the Web, you would imagine there would be similar enthusiasm for the topic among sleep experts. But, oddly, that's not the case. Some sleep textbooks don't mention sleep talking at all. Kryger's *Principles and Practice of Sleep Medicine* is a hefty tome found on the desks of most sleep doctors and researchers around the world. It weighs in at 1,552 pages and yet sleep talking only manages to command a paltry thirty-six lines (scattered across five different chapters) and a couple of passing mentions. That's not to say that there hasn't been a fair bit of research into the field, but we know very little more about it now than we did thirty years ago. Perhaps this is because it is common, and is seen more as a source of amusement or mild annoyance than as a serious problem. Indeed, there seems to be growing doubt in the sleep science community about whether to consider it a disorder at all, as evidenced by its status in The International Classification of Sleep Disorders. In the first edition, sleep talking made the grade as a disorder and was listed under "SleepWake Transition Disorders." But by the time the second edition came out people weren't so sure. It has now been relegated to "Isolated Symptoms, Apparently Normal Variants and Unresolved Issues."

So what is sleep talking, what causes it, and what does Sleep Talkin' Man tell us about Adam's unconscious? To really get to grips with this we need to take a closer look at sleep.

For thousands of years we assumed that sleep was a single thing. You were either awake or asleep. But with the invention of the EEG we discovered that sleep was much more complicated and researchers have spent the last sixty years trying to understand what is really going on, with only partial success.

Sleep is actually made up of several different types of sleep, or sleep stages, which alternate through the night in ninety-minute cycles. Stage One sleep is very light sleep and is usually a brief transitional stage we pass through between being awake and being "properly" asleep. That twilight, half-asleep, half-awake sensation you get just as you're drifting off to sleep is probably a sign that you are in Stage One sleep.

Stage Two sleep is also pretty light, but this is unequivocally sleep and you spend about half the night in Stage Two. What may surprise you is that you don't stop thinking throughout this stage. There appears to be mental activity of one type or another for much of the time you are in Stage Two of sleep, and there have even been dreams reported in Stage Two. So it is not unexpected that where sleep talking occurs it often comes out of this stage of sleep.

Stages Three and Four are lumped together and referred to as Slow Wave Sleep. Slow Wave Sleep is the deepest stage of sleep; in fact it is the most unconscious you will ever be without being under anaesthetic or knocked out by a blow to the head. Yet even in this deeply sedated state there is evidence that your brain doesn't turn off completely. Counterintuitively, in some people this can be a very active time, because Slow Wave Sleep is when sleepwalking occurs! We don't know for sure why people sleepwalk, but the really deep sleep of Slow Wave Sleep is vital to this process. What seems to happen is that the somnambulist (sleepwalker) somehow partially wakes up from Slow Wave Sleep. Those parts of the brain that allow them to see, move, eat, unlock doors, drive cars, dance, or have sex wake up. But those parts of the brain that control purposeful behavior, exercise sound judgement, strive for goals or restrain the person from doing inappropriate things remain deeply asleep, blissfully unaware of what the rest of the brain is getting the body to do. Thus the somnambulist may eat very skilfully with a knife and fork, but what they eat may be a box of Kleenex. If you can sleep dance then you can sleep sing and so some sleep talkers do their talking when in Slow Wave Sleep.

The final sleep stage is Rapid Eye Movement Sleep or REM, and this is the strangest sleep stage of all. REM is where you do most of your vivid dreaming. You spend a quarter of the night in this state of intricate, intense, and often bizarre hallucinations. But it gets even stranger. In REM Sleep, we temporarily become cold blooded. As mammals, we keep much the same body temperature no matter what the outside temperature is. But in REM we lose this ability—if the room temperature goes up, so does our body temperature; if the room temperature goes down, our body temperature drops too.

But what is most relevant to sleep talking is another odd feature of REM Sleep. When you dream that you are running down the road, that part of your brain that makes your legs run is actually firing. Now, clearly running in bed is not a great idea, so a group of nerves at the base of your brain set up a sort of road block that stops those nerve impulses reaching your muscles. As a result, when you are in REM Sleep you are literally paralyzed. Only your diaphragm, your eye muscles, and middle ear muscles are able to move. For this reason, sleep talking is not as common during the dreams of REM Sleep as one might have expected. Some sleep talking does come out of REM Sleep, but if it weren't for the sleep paralysis it would happen almost every time we dreamed.

So which stage of sleep does Sleep Talkin' Man dwell in? The biggest clue is in the timing. Although the various sleep stages alternate in cycles, these cycles change gradually across the night. The first couple of cycles have lots of Slow Wave Sleep but very little REM and Stage Two, while the later cycles are almost entirely REM and Stage Two Sleep. Sleep Talkin' Man appears far more frequently in the second half of the night, between 4:30 and 6:30 a.m., so this pretty much rules out Slow Wave Sleep. Also, Sleep Talkin' Man does sometimes respond to external stimuli, like sounds. This is unlikely in deep Slow Wave Sleep, but quite common in the lighter Stage Two and REM stages.

From here on we are relying on a bit of educated guess-work. As Adam doesn't recall having dreams if he wakes up at the tail end of a Sleep Talkin' Man utterance, it seems unlikely that Sleep Talkin' Man is in REM Sleep. So on balance, Sleep Talkin' Man seems to be a product of Adam's Stage Two Sleep.

We now have an idea where STM comes from, but haven't yet worked out what it all means. Is this normal or should Adam and Karen be really worried?

The first thing to say is that sleep talking is very common, especially in children. In fact, sleep talking has been reported in up to half of young children! It is actually quite possible that the vast majority of us sleep talk at some point in our lives with varying levels of comprehensibility, emotion, and frequency. This creates a problem for scientists. When they want to compare people who sleep talk and people who don't, they have quite a time finding people who have never sleep talked. Some people who swear blind that they never sleep talk can turn out to be quite prolific sleep talkers when they are recorded at night.

Adam's story is a bit different. Other than the odd mumble at college, he seems to have been a pretty silent sleeper until the age of thirty-four. Adult onset sleep talking is much rarer, though by no means unheard of. Nor is Sleep Talkin' Man's impressive canon of work unprecedented. Sleep talking several times a night, most nights has been reported in numerous scientific studies of prolific sleep talkers. Why some people sleep talk more than others isn't known for sure but there is often a genetic component to sleep talking so it is quite possible that we may one day have Son of Sleep Talkin' Man!

So sleep talking is probably normal and Adam's case is not as unusual as one might imagine. But, the big question everyone is asking about Sleep Talkin' Man is what he says about Adam. Does the content of Sleep Talkin' Man's remarks reveal something dark and sinister about Adam's character?

The first point to consider is whether Sleep Talkin' Man is simply accurately reporting what Adam is thinking. Unfortunately we can never know for sure, because Adam rarely remembers what he was thinking when he wakes from a Sleep Talkin' Man episode. Sometimes sleeptalkers say exactly what they are thinking, sometimes they say things that are related to what they are thinking, and sometimes what they say has no relationship to what was actually going on in their heads. Adam and Karen have been able to identify numerous times when Sleep Talkin' Man appears to be referring to something that has been going on in their lives. For example, elephants appeared prominently in Sleep Talkin' Man's musings when they were on honeymoon in an elephant sanctuary. It stands to reason that Adam would have been thinking or dreaming about elephants in his sleep and that this came out in his sleep talking. But whether what Adam was thinking about the elephants and what Sleep Talkin' Man was saying about them is the same thing we may never know.

It seems that when Sleep Talkin' Man appears, some parts of Adam's brain are fully functioning while others are not. His speech and language generating circuits are clearly working, but perhaps those parts of the brain that exercise judgement and tact, or govern logic are still fast asleep. Thus Sleep Talkin' Man speaks with perfect grammar but what he talks about is irreverent, uninhibited, and full of bizarre connections that the waking mind, constrained by the need to be orderly and logical, could never generate.

Some people worry that Sleep Talkin' Man may be an indication of insanity. It is true that psychiatric disorders are about twice as common in adults who are frequent sleep talkers as compared to non–sleep talkers. Neverthe-less, the vast majority of adult frequent sleep talkers have no psychiatric problems whatsoever. Remember, even the sanest people in the world have crazy dreams on a nightly basis and we probably all have bizarre and disjointed thoughts when we sleep. It's just that most of us aren't generous enough to share these thoughts with the world.

Sleep is a strange and mysterious state. It is almost inevitable that it will sometimes create something as strange and mysterious as Sleep Talkin' Man.

Dr. Hugh Selsick is a psychiatrist with a special interest in sleep disorders. He is Chair of the Sleep Group in the Royal College of Psychiatrists, and is Lead Clinician of the Insomnia Clinic at the Royal London Hospital for Integrated Medicine/ University College London Hospitals

**Legs time!
Everybody get your legs!**

> "Awesomeness now has a name.
> Let me introduce myself."

Adam made it through thirty-six years of life without the company of his alter ego. So, why did Sleep Talkin' Man suddenly emerge out of Adam's subconscious depths that cold February night, and what purpose does he serve?

Many sleep specialists hypothesize that sleep talking is linked to stress. Thinking back to when STM first showed his face (or mouth, as the case may be), that theory sounds about right to me. Back in February 2009, Adam and I were under a number of tremendous stresses. Allow me to paint you a picture:

A year earlier (as soon as we got engaged), we had applied for a visa for Adam to come live in the States with me. His immigrant visa had been rejected in May (that story is another whole book right there), and we had appealed. By February, we had already spent over EIGHT MONTHS waiting to hear the results of our appeal. During that entire time, our life was in limbo, waiting for the U.S. government to determine on what continent we would settle down together. Because Adam could not enter the States while his appeal was in process, our only option to be together was for me to come to London. However, I had an apartment and Molly the little beagle in New York. So all that time, I commuted back and forth between my two lives—six weeks in London, three in New York. It was painful to spend so much time separated, when we already felt like we had lost so many years together. And it is difficult to communicate how nerve-racking it is to live that way for so long, waiting on a faceless agency to decide something so major in your life, losing control over something as fundamental as on what continent you can choose to make your life. It sucked.

Of course, in order to be able to spend two-thirds of my time with my fiancé in the United Kingdom, I had to quit my job. During this whole time my status in the United Kingdom was that of a visitor, so I was unable to do any work except for bits of freelance that I picked up from the States. Meanwhile, Adam had been looking desperately, but without success, for a job in his industry for most of a year. The economy had tanked, and advertising agencies were on a hiring freeze. Although it was wonderful that we got to spend so many of our days together whenever I was in London—and I would happily do that for the rest of my life—it wasn't a situation that was very good for our bank accounts, or for Adam's self-esteem. It really sucked.

With no real income, and no notion of when we would get an answer and where we would end up, signing a lease on an apartment was out of the question. We spent those months moving back and forth between Adam's parents' home and a tiny little studio apartment we could sublet for a few weeks at a time in the East End, with the occasional heavenly stint house-sitting whenever Adam's brother's family was on vacation. As lovely and generous and wonderful as Adam's parents were, and as kind as his brother was to lend us his home, we felt like nomads. We simply had no place to settle in and make our home. It royally sucked.

Finally, just in case all of this wasn't anxiety-producing enough, we spent much of this period immersed in an acrimonious legal battle concerning visitation with Adam's children. Those of you who have been through it know how much THAT sucked.

So I think I've made my point: lots of stuff kind of sucked, and it was probably a fair amount more stress than the average first world couple has to contend with in the first two years of their relationship. Given all of that, one can easily understand how the acute stressors in Adam's life might have overwhelmed his psyche enough to cause STM to bubble to the surface. And yet, I don't believe that's the whole story. I don't think this strain alone explains why Adam began talking in his sleep. You see, Adam had been through even greater stress and hardship before in his adult life—a difficult marriage followed by a bitter divorce can be truly soul-shattering for a sensitive spirit—without making a single late-night peep. There must have been some other contributing factor.

Perhaps it was this: in spite of all of the difficulties that we were facing, Adam was the happiest he had ever been up until that point in his life. Yes, we were struggling with our circumstances, yes, our future was hanging in limbo, but we were thrilled to be in each other's company and starting our life together, and every day was a joy.

So, perhaps STM's emergence can be attributed to a combination of these two things: Adam's intense anxieties and frustrations had his subconscious running in hyperdrive, and because he felt secure and accepted, he felt the freedom to work it all out of his system through the catharsis of sleep talking. Considering how even-keeled and composed Adam remained in the face of every adversity we faced (far more than I did), that theory adds up in my book. So, you see, we think of STM not as a revelation of the dark abyss of Adam's psyche, but rather as a sign of Adam's mental health. STM is simply a processor—negative energy is metabolized into nightly nuggets of comedy gold. Ironic? Perhaps. But this way, everybody wins.

Thus, when I settle into bed each night, I know that at some point I may be awakened by Adam's subconscious playing out whatever bizarre catharsis will help to keep his day free and easy. And, more importantly, I know that in the morning, Adam will be lying next to me, ready to share a moment's laughter, and be the best husband I could ever hope for.

Glossary

Bog – *n* Toilet, vulgarly. Used as verb in "bog off," the order to go away. Similar to "fuck off," but lighter on the profanity.

Bollocks – *n* Slang for testicles. And, similarly to "balls" in the States, may be used as an exclamation in a moment of pain or frustration. For example, "BOLLOCKS!! I slammed my hand in the car door. OWWWWwwwww."

Bugger – *v* Slang for the act of one performing anal intercourse on another. Can also be used as an exclamation of frustration, "BUGGER! I accidentally deleted that file!" or be used in the phrase "bugger off" to mean "go away."

CV – *n* Abbreviation for curriculum vitae, a synonym for résumé. In the States, this term is gnerally used only in academic circles. In the United Kingdom, it is used almost exclusively.

Fairy lights – *n* a long string of tiny bulbs used for decoration. In the States we generally referred to them as Christmas lights.

Fiddling – *v* Slang, for committing a pedophilic act. Generally used in the phrase "kiddie fiddling."

Five-a-day – *n* Daily ration of vegetables recommended by the British government.

The fucking bollocks – *adj* Used to describe something that is fantastic. As in, "This cheesecake is the fucking bollocks!" One can also use "the dog's bollocks" in the same way, although I'm not sure what is so great about a dog's testicles.

Knickers – *n* Ladies underwear. Used when Americans might say "panties."

Knob-end – *n* Literally, the head of a penis. Used to describe someone who, in the States, might be called a "dickhead."

Jumper – *n* The item of clothing that in the United States would be referred to as a "sweater."

Pants – *n* Underwear. This is ALWAYS the meaning of "pants" in the United Kingdom. If you mean to refer to the article of clothing that people wear to cover their legs, be sure to say "trousers." This word has been the most difficult for me to adjust to since I immigrated. More than once, I have called across a crowded shop, "Adam, check if they have any good pants for you!"

Pissed – *adj* Drunk. If you mean to describe someone who is angry, "pissed off" still works as it does in the States.

Spot - *n* Slang for pimple. The equivalent of "zit" in the States.

Swede – *n* The British term for the yellow turnip that we in the States refer to as a rutabaga.

Tosser – *n* See wanker. "Tossbag" can be used the same way.

Wanker – *n* Someone who acts like an idiot or asshole. While "to wank" means "to masturbate," "wanker" is rarely used for anything besides a term of general disrespect.